This Accursed Land

Lennard Bickel was a former journalist for the Australian Broadcasting Commission and the Australian newspaper. He published books including *Facing Starvation* and *Shackleton's Forgotten Men*.

The Attempted Land

LENNARD BICKEL

THIS ACCURSED LAND

CANELOHISTORY

First published in the United Kingdom in 1977 by MacMillan London
Limited

This edition published in the United Kingdom in 2021 by

Canelo
31 Helen Road
Oxford OX2 0DF
United Kingdom

Copyright © Lennard Bickel, 1977

The moral right of Lennard Bickel to be identified as the creator of this
work has been asserted in accordance with the Copyright, Designs and
Patents Act, 1988.

All rights reserved. No part of this publication may be reproduced or
transmitted in any form or by any means, electronic or mechanical,
including photocopy, recording, or any information storage and retrieval
system, without permission in writing from the publisher.

A CIP catalogue record for this book is available from the British Library.

Print ISBN 978 1 80032 549 4
Ebook ISBN 978 1 80032 548 7

Look for more great books at www.canelo.co

Printed and bound in Great Britain by Clays Ltd, Elcograf S.p.A.

To

FRANCES CRAIGHEAD

who opened a door

Foreword

The heroic period of Antarctic exploration occurred in the score of years at the beginning of this century... a surprisingly short time ago when you think of how long man has been exploring his planet.

It was in that time that men first commenced their resounding battles, with all the incredible discomforts and hardships of the southern continent, and the period was dominated by the names of Amundsen, Scott, and Shackleton – and mighty men they were, too.

But, there were others like these great heroes who were undertaking comparable feats of courage and leadership, who never achieved quite the same stature in the public mind because their objectives did not include that magic goal – the South Pole. Outstanding among these men were two Australians, middle-aged Professor Edgeworth David and Dr Douglas Mawson (later Sir Douglas). Their 1908 manhauling sledge journey to the South Magnetic Pole along with the Sydney medico, Dr Alistair MacKay, was a tremendous undertaking. It was on the return journey of some hundreds of miles of desperate travel back to the sea coast that Mawson showed to the full the strength, vitality, and leadership that were so pronounced a part of his character.

But Mawson had an even greater test ahead of him. From 1911 to 1913 – at the same time as the Scott tragedy was coming to its sad conclusion – Mawson led his Australasian Antarctic Expedition into the unknown country west of Cape Adare. It was an expedition that carried out a notable amount of scientific research, which earned it a place among the great scientific expeditions of its day.

However, the most dramatic events had nothing to do with science – they involved a man's battle against most appalling natural obstacles... Mawson's own fight with death, as it is related in this book.

Some 320 miles east of Commonwealth Bay (which Mawson discovered) a companion crashed into a deep crevasse, with their tent, all the man food except one week's sledging rations, all the dog food, and most of their equipment. With his other companion – who was also doomed – Mawson's journey back to winter quarters was a terrible journey. It became what is probably the greatest story of lone survival in Polar exploration.

That he survived at all was due only to his tremendous spirit and determination. Mawson earned his place of honour as one of the great men of the Antarctic.

SIR EDMUND HILLARY
Sydney
1976

Prelude

The Two Tents

The tents were a thousand miles apart. Between them was an awesome waste of ice-clad mountain ranges, huge glacial valleys, wide upland snow plains swept by polar wind. Yet – separated as they were – they were tightly linked in the saga of the Antarctic.

Three men lay in each tent. Both tents were cut from green canvas, but the one in the south-east, pitched on the vast floating wasteland of the Ross Ice Shelf could not be seen. It was completely covered by the compacted snow of a long winter. It was quiet and still inside that tent. Not so in the shelter on a high ice plateau in the north-west. There, three men stretched out in reindeer-skin sleeping-bags and listened to the howl of a ferocious blizzard, which slapped and banged the canvas cover against the bamboo poles. It was so noisy they could only communicate by shouting – but, by now there was little left to say.

For two days they had huddled down trying to keep warm, trapped on the exposed plain while the winds from the polar plateau rushed unimpeded across the bottom of the world and filled the air with racing drift and stinging ice crystals. In that dense gloom they had neither night nor day and the hours were a long painful crawl. The winds were so strong that they could not ignite their kerosene-fed primus stove, and unable to melt snow for drinking water or to heat food, they waited in their bags and endured thirst, and eased hunger by nibbling hard-tack biscuits and small sticks of plain chocolate, listening always for some sign of a lessening in the gale.

Each one of these three men might easily have been in the other tent. Most certainly the leader of this trio and leader of the whole expedition, Dr Douglas Mawson, would have been had he not resisted strong pressure to abandon his own plans. But, for his two companions, he knew, it was a second choice. Both of them had originally sought to join the other party and – missing selection narrowly – had been warmly recommended to Mawson. He had found them first-class men, ideally suited to their allotted tasks, and so had been glad to appoint them.

Mawson was already an Antarctic veteran at the age of thirty, tall and powerful, a commanding leader and experienced geologist. He was encamped with two men from different countries and with different backgrounds. Lieutenant Belgrave Edward Sutton Ninnis, of the British Royal Fusiliers, scion of an old Cornish tinmining family, son of a London doctor, was unassuming, patient and loyal, and 'good with animals'.

Dr Xavier Guillaume Mertz, a German-speaking law graduate from the Swiss city of Basel, was at twenty-eight years of age a ski-champion and fine mountaineer. Ninnis and Mertz had become firm friends since they joined this expedition in London as joint handlers of the thirty Greenland husky dogs and crossed the world together on a long, slow voyage, tending and nursing the animals on the open deck. Now the dogs were snuggled down in the snow with only their muzzles showing. Both dogs and men waited for the wind to die and the air to clear when they could set their faces eastward and move into the unknown wilderness. All three men were strong, determined, conditioned by hard work in the long dark months of winter, impatient to open their journey of discovery.

Into the afternoon of that second day the gale blasted from the hinterland and the rivers of snow filled their world with white, frenzied chaos while they lay waiting under the rocking tent, cold, thirsty, hungry.

Over the tent in the south-east the afternoon was clear, calm and still, and a low sun bounced its vicious glare across the 250,000 square miles of the white wastes of the Ross Ice Shelf. It was a day to be engraved in polar history, a day of sad discovery. Ten men, with two ponies and two dog teams, had plodded into the glare since early that morning – for almost twelve hours they had trudged southward.

The searching ended when they found the cone of snow and beneath it the tent that had gone to the South Pole. Under the snow's weight the tent walls sagged over the bodies of Captain Robert Falcon Scott, Dr E. A. Wilson, and Lieutenant Henry Bowers. Diaries with the bodies told the now classic tale of the crushing blow of finding Amundsen had reached the Pole weeks before them, of the struggle back in which two of the party of five had perished. Now, here were the three of them dead only eleven miles from food and fuel at One Ton Depot.

The bamboo poles of the tent were pulled away; the green canvas became a shroud, and a cairn of snow was built over them. There was a short burial service and the dead were left to become integral with the ice. The historian of the Scott expedition, Apsley Cherry-Garrard, described it as a tomb that kings might envy – and a *fitting mausoleum for the last of the great geographical explorers.* (Author's italics.)

He could not know that another giant of Antarctic exploration was to his north-west at that moment, waiting with two companions to commence his share of geographical and scientific exploration more extensive than was done by Scott's larger and better funded mission. That this man, Douglas Mawson, would add more territory to the maps of the world's sixth continent than any other man of his time.

But, no man on the southern ice that November day, 1912, could foresee how the journey into the eastern wilderness would become another bitter chapter in the Antarctic saga. For this terrible march was to be the second epic in that year to inflict great suffering, bring death to brave men, impose an ordeal grimmer than that that took the lives of Scott and his polar party, and end in an incomparable struggle for survival in one of the most savage environments on earth.

The Cruel Continent

The Antarctic is alien land. Caught in a frozen grip, desolate and barren, hostile to life, it is a lost continent at the bottom of our world, now smothered under the greatest ice-shield known.

This solid ice-cap is an immensity. Three miles thick in places, with a mean, overall thickness of a mile and a quarter, it blankets almost 6 million square miles of the Earth's surface in southern summer – a region bigger than Europe and the United States combined. In winter, when the sea freezes over, the area inundated with ice can be doubled. On the mainland, from the high and remote solitude of the South Pole and the so-called Pole of Inaccessibility, down to the coastal fringes, only two per cent of the rocky land is able to break free of the frozen mantle.

Far inland, the ice-plains rise more than 12,000 feet above sea level, and only the frost-rimed peaks of the mightiest mountains can pierce this frigid shield. Cloaking, submerging the great ranges, the ice makes the Antarctic the highest overall continent on Earth and exerts an influence and an impact on the world's weather to an extent not yet fully understood.

This colossal canopy is of enormous weight and has prodigious power. Some 7.5 million cubic miles of ice puts the pressure of 24,000 million million tons on the buried continent and crushes the mountains, plains and valleys back into the Earth's mantle. The contour of the planet is thus dramatically flattened at the South Pole. Fortunately for mankind, the outward thrust of this gigantic mass is slow. The melting of the ice in the oceans is minimal. If, by some cataclysm, all this Antarctic ice was to melt, the resultant flood of 5.4 million million million gallons of fresh water would raise the sea level of the Earth above the decking of the highest bridges, including San Francisco and Sydney, and have catastrophic impact on many forms of marine life.

However, most of the southern ice is glacial – ever on the move outward and downward from those distant polar plains to the great oceans that ring the globe at those latitudes. Its size and altitude, and cold, frame a climate unique to Earth, a climate that breeds the worst winds known; for, from the rarefied highlands, the intensely chill, heavy air falls down the slopes, shooting through the glacial valleys and the shoulders of the ranges, gathering momentum with gravitational impetus, to launch unparalleled onslaughts on the coastal plains, to lash shore waters into turmoil and the fields of pack-ice into furious upheaval.

These deeply-chilled winds are chisels of air that carve and curve the surface of the ice as though it was water, shaping it into corrugations explorers know as sastrugi, frozen waves that, beneath blown drift, lurk as traps for unwary feet. They are but ripples on the face of an ice canopy that effectively masks the nature of the frozen, buried lands. Only fragmentary clues escape; debris brought down by the grinding beds of glaciers yield traces in the species of rocks of gold, coal, oil, uranium – mineral treasures inaccessible under the most massive ice-cap of all.

The land was not always smothered. Hundreds of millions of years ago, it seems, it was part of a vast land mass that, under stress from movement in the planet's crustal plates, was sundered into separate continents. One magnificent expanse of wide grassy plains, rich jungles and high chains of peaks with gleaming lakes slipped westward to form the great American continent… and hanging on its toe, held by an umbilical cord of tough gneiss rock, the doomed land slid southward.

Slowly, through uncounted millennia, the land inched across the bed of the young, dark seas. Stresses in the Earth's tectonic plates unleashed the fury of enormous volcanoes to pock-mark the face of the southerning land and caused earthquakes, which thrust mountain peaks 16,000 feet above the sea until, at last, the land reached the end of the earth.

Once out of the sun's direct warmth the land was buried under the gigantic frozen overburden and enclosed behind seas of floating ice and towering ice ramparts.

Today, the sixth continent sprawls over the bottom of our world, shaped somewhat like a white frozen fist. Out of this clenched hand one rocky finger of territory breaks free and reveals the geologic link with the majestic mountain chain that is the backbone to the whole American continent. Up through rugged Graham Land and the South Shetlands the rock finger stabs north, dips beneath the stormy waters of the Drake Passage toward Cape Horn.

There are more bonds than geology. There were links in animal, marine, insect and plant life with the ancestral terrain, with America, Africa, Australia – even Asia. Yet, such links perished in the cold and winds; but not all were totally obliterated. Today the plundered colonies of whales, seals, the penguins, sea leopards, and the myriad birds that prey on marine life still haunt the coasts and fringe islands in southern summer. Inland, there are mere traces of long-gone verdancy. A glacier cuts the thin face of a coal seam laid down in prehistory; exposed are the fossil leaves of an extinct tropical forest; glacial moraines bring down pieces of petrified tree trunks or fragments of bone from the mangled skeleton of some animal of antiquity.

On the frozen continent extermination of life forms was inevitable. And still is. Winter comes when Earth is at the most distant point from the solar centre, and summer brings only angled glimpses of sunlight, for a few weeks. Elevated as it is above the warmer air at sea level, it is much colder than the Arctic north. High on the uplands, round the desolate polar plateau on those 12,000-feet-high platforms a cold of rare intensity freezes the air into a crystallised white-grey mist, a shroud that drops a deathly touch over the frosted peaks and plains, which falls down into the glacial valleys toward the coast.

8

It is cold that kills. It is the coldest cold on Earth. It makes the air so heavy it falls across the frozen plains with increasing speed, hardening the ice to brittle rigidity so that the glaciers – among them the greatest on the planet[1] – rend and fracture as they twist their way to the sea and are riven into deep fissures, jagged, winding cracks, called crevasses, reaching to bedrock. The frozen rivers break their backs and burst with enormous explosions, like a bombardment by massed artillery… when it can be heard above the roar and the boom of the gales.

Cold makes the Antarctic alien – but winds make it more deadly. The worst, and most dangerous, are katabatic winds, flying rivers of air, cold and heavy, falling down the frozen slopes from the polar plateaux and increasing in speed, with gravity, to assault those parts of the coast where they find outlet. They reach gusts of above 200 miles an hour and can blow consistently for days and not drop their force below 80 miles for many hours on end. Such winds lift gravel and hurl rocks and heavy objects out to sea – they blow men from their feet and their breath encases eyes, nostrils, mouths in ice. They are the worst winds in the world, a greater menace than cold. Born in rare high solitudes, they pick up snowflakes, ice crystals and frozen pellets, compacted like hail, all of which, blown in the wind, become abrasive material that

---

[1] The Lambert Glacier is the biggest on Earth; it rises in 100,000 square miles of ice surrounding the Prince Charles Mountains, south of the Australian base of Mawson. Some 500 miles long, 200 miles wide in places, it feeds the big Amery Ice Shelf.

can polish rough metal to brilliant sheen and scour the wood from between the grains when they are left exposed for a winter. Cold and wind can reach the sheltered parts of a man's body and cause deadly frostbite, adding to his peril.

Men are always surrounded by danger and the hazards can change, constantly. The canopy of ice is ever on the move, the glaciers strain, shift and break; what may be safe today will be perilous tomorrow. The ice has a life cycle; each snowflake that falls on the polar plateau may form part of the outward flowing mantle and eventually reach the sea in one of the frozen barriers, a shelf, or a glacier tongue. Once there, it may break off as part of an immense iceberg, a floating island of ice that the wind will carry north, to beyond the Antarctic Circle, above sixty degrees South, where the chilled southern waters sink below the warmer tides of the Atlantic, Indian, Southern and Pacific Oceans: where the sun will melt the ice and lift the moisture back into the atmosphere, perhaps again to fall as snow over the South Pole.

The Assault

The hostility of the sixth continent overflows its frozen borders. Outside the walls of sheer ice are savage defences and seas that can be treacherously beautiful, and which held back exploration until two centuries ago. Men in wooden sailing ships were spurred south in the Middle Ages to discover the mysterious southland, which cartographers believed existed to balance the terrain of the northern half of the world.

But these intrepid sailors were beaten back by the first lines of the Antarctic defences. Tempestuous winds sweep those vast open seas, and calm only brings dense fog and white mists that shroud immense floating death-traps – islands of ice, fields of growlers and jostling pack, tossing floes that can overnight – in an hour even – freeze over and squeeze a captured ship to matchwood. And silently the massive bergs slide through the sea with submerged projections that can sink the greatest vessel with a sideways graze.

Nevertheless, penetration by man into this hostile region was opened by wooden sailing ship, by the aptly named British vessel, *Resolution*. Conned by the intrepid Captain James Cook, it carried men for the first time inside the Antarctic Circle in late 1773, and, in the following January, thrust even further south. Forcing his ice-coated ship into dangerous waters, creeping through fogs and mists, dodging bergs and pack-floes, Cook reached beyond seventy-one degrees south. Then, in waters now called Amundsen Sea, offshore from territory we know as Byrd Land, he could go no further south. He was faced with towering walls of ice. By dead reckoning he calculated he was then some 1,250 miles from the geographical South Pole, and he believed he was very close to the mainland – but the Antarctic light tricked him. Cook could see south, beyond the icy ramparts, to where white-crested mountains soared into a distant sky, and understood he was close to an ice-bound land. He was most certainly the first man to glimpse the peaks of long-lost mountains – but he could not have seen them by direct line of sight. From his most southerly point the closest peaks, topped by the 8,000-feet-high Mount Murphy, were near to 300 miles distant... beyond the curvature of the earth.

Almost certainly Cook was a victim of the Antarctic mirage in which layers of cold air of differing temperature can reflect a landscape into the sky so it can be seen far beyond the horizon. It is a deception now well-known to southern polar travellers. But, whatever the men aboard the *Resolution* thought they saw that day, they were certainly touched with awe and insight. Cook made a log entry:

'It was indeed my opinion, as well as the opinion of most on board, that this ice extended quite to the Pole, or perhaps joins some land to which it has been fixed from the creation.'

And from there even Cook was glad to turn north. He went on to circumnavigate the continent, sailing right round the globe before heading north with firm convictions on the Antarctic: the hostile southern land was inaccessible and no man could ever penetrate it. 'I can be so bold to say no man will venture further south than I have done, and that the lands to the south will never be explored.' For him the sixth continent was 'doomed by nature never once to feel the warmth of the sun's rays, but to be buried in everlasting snow and ice.'

Cook's warning was not conclusive. Drawn south by his reports of teeming coastal life, predatory whalers and sealing boats ploughed through 12,000 miles of seaways inside the Antarctic Circle, profit-taking in carnage among the whales, seals, and migratory birds that breed on the coasts, countless islands and rocky southern capes. The men found new islands, saw distant white land, and, when profits were not high enough, turned to the slaughter of the timid, teeming penguins for the miserable amounts of fat and oil their bodies contain.

Cook's prophecy was to be proved false within fifty years. In 1823, James Weddell, of Britain, sailed below America to seventy-four degrees south. Seven years later, one of the captains of the firm of Enderby Brothers, of London – John Biscoe – sighted land south of Africa: another Enderby captain discovered a group of islands and named them after himself, the Balleny Islands.

By the mid-nineteenth century, exploratory, rather than exploitative, expeditions entered the southern lists. Bellinghausen, of Russia, repeated Cook's circumnavigation of the continent and found an island or two south of America; the Frenchman, Jules Sébastien Cesar Dumont d'Urville, while landing on an island, caught a glimpse of an ice-girt rocky cape and, claiming that sliver of land – directly south of Adelaide – for France, called it Adelie Land, after his wife, a name far too graceful for such a harsh land.

Like infrequent mice nibbling at the edge of a vast, chilled cheese, the expeditions came and went through the rest of the nineteenth century. Sir James Ross with his Royal Navy squadron had probed the Ross Sea region; Captain James Wilkes with a US Navy flotilla swept the sea and was deceived into marking land into his maps that others later sailed over. But a century and a quarter went by after Cook's furthest journey south before any man set foot on the Antarctic mainland. A whaling venture in 1895, led by an Australian, Henry Bull, with Captain Leonard Kristerson, probed along the ice barrier of the Ross Ice Shelf, found a rare gravel beach, and got ashore for a few hours. This first landing site is due south of the New Zealand port of Dunedin. It was subsequently named Cape Adare, and both landing and site were primers for the invasion that ensued.

In 1897, Adriane de Gerlach of Belgium took the *Belgica* south to a harrowing winter. His ship was trapped in ice and drifted for a whole year inside the Antarctic

Circle, and the first mate of that ship was a Norwegian, named Roald Amundsen.

By the time the *Belgica* broke free, plans were afoot for the first wintering party to go ashore on the mainland at Cape Adare. Out of Melbourne, the Southern Cross Expedition – named after the ship – sailed south in late 1898 under the command of Carsten Borchgrevink, and funded by the British publishing tycoon, Sir George Newnes. Adventurer, explorer, whaling man, and a very practical soul, Borchgrevink was a Norwegian-born resident of Australia. He carried enterprise into the southern continent.

Landing safely on the beach he erected the continent's first shelter – a prefabricated hut – and 124 years after Cook's prophecy on the 'doomed' land, the air echoed with the sound of barking dogs and men's voices. Borchgrevink introduced the husky team and the sledge to the continent, though he was penned in and could not travel very far because of ice conditions. He also lodged another 'first': he buried the Antarctic's first victim – Nicolai Hanson, a taxidermist whose death appeared to have all the signs of incipient scurvy.

In this same year, national pride and interest was stirred in European capitals; there was a meeting in London of the International Geographical Congress from which came the framework of a multi-national assault of separate expeditions, and which also sealed the destiny of an obscure young torpedo officer in the Royal Navy – Captain Robert Falcon Scott – and set the pattern of exploration for the coming decades. Expeditions were launched from Sweden and Germany. The Swedish ship was trapped in ice and the men spent two winters on Snow Hill Island, in the American Quadrant, while the Germans, under Erich von Drygalski, were also ice-bound south of Western Australia at a place they called Gauss Berg, near the vast sheet of floating ice that was to be named for a young naval reserve officer then serving with Scott – Ernest Shackleton.

Scott's persistence overshadowed the German and Swedish failures. He pioneered long-distance sledge travel. He first landed at Cape Adare and then went on to find anchorage for his ship *Discovery* in McMurdo Sound where Ross had named a soaring active volcano after one of his ships – Mount Erebus – and a dead volcano after another ship – Mount Terror. For Scott these two peaks stood at the portal of what he was to make the highway to the South Pole. With much to learn – and suffer – he took the rigid formality of the Royal Navy onto the ice; officers were officers and men were always men, and the mess was run like a wardroom. On this regime in this most southern human habitation, he developed the sledging disciplines and the hard rigorous approach on which his great reputation was to be built.

Through the months of 1902, Scott imposed routines for sledging and camping, and in November set off to the south across the Ross Ice Shelf for the first human invasion into the Antarctic interior. He had two companions – Dr E. A. Wilson, doctor and artist, and a rugged and ambitious reserve officer, Lieutenant Ernest Shackleton. They took nineteen dogs and a carefully allotted food ration. The wind, cold, and terrible terrain wore down their condition; the dogs died one by one; the men – especially Shackleton – suffered from scurvy, but they fought their way on a most harrowing, gruelling journey to a latitude of eighty-two degrees south, eight degrees from the bottom of the world. Scott was then compelled to turn back. With cold, malnutrition, wind carving their faces, haggard from endless toil, they came back to safety by the barest margin. They carried the sternest warning on the savage nature of the inland territory.

Shackleton was very ill; and his illness was to have great impact on all that followed. Scott did not think he would survive another winter in the Antarctic, so sent him home to recuperate aboard the *Morning* when it came south with fresh stores and fuel. Shackleton was feted, lionised in England; one of three Englishmen to get closest to the South Pole and the first to be home to tell the tale, he responded to fame – and opportunity. Glowing with national pride and ambition he became entrepreneur-explorer and soon enlisted the rich and the influential to his openly declared objective – to reach the geographical South Pole.

Lectures, talks, articles, pleading, persuading, he peddled the Pole as a national objective until finally he had his own expedition staffed and equipped. His men named him 'The Boss', a title he loved, and which suited him. Flaunting his battle-cry, 'First to the Pole', he went south in August 1907 in the *Nimrod*, a former sealing ship.

Shackleton's personal magnetism was striking, his sense of promotion remarkable. The Pole was the goal, science a side issue; but, still, he astutely added Australian and New Zealand scientists to his staff, thus winning cash support from both governments. Among those enlisted was fifty-year-old professor of geology at Sydney University, a deceptively gentle man and lovable tutor known as 'Tweddy' – Dr T. W. Edgeworth David. David thought Shackleton should have a magnetician, cartographer and surveyor, and to fill all three posts, he recruited a former outstanding student, then a lecturer in geology at Adelaide University, Dr Douglas Mawson, who, at the time of the invitation, was investigating an outcrop in South Australia that proved to be the first discovery of uranium in the land.

Like Cook, Mawson was born in Yorkshire. When he was two years of age his parents – whose family had lived there for generations – migrated to Australia, with Douglas and his brother William. When Shackleton's invitation arrived, Mawson was twenty-six years old, a man of exceptional physical strength, gifted with an adventuring spirit, a scientist's mind, and a natural bent for leadership. Hardened, weather-browned by many expeditions into Australia's outback, six feet three in his socks, blue-eyed, Mawson joined 'Tweddy' and went south with Shackleton to face a world of ice. Aboard the *Nimrod* he met two men who were to share his future; the first mate, John King Davis – a glum, red-haired Irish sailor, soon to gain the nickname of 'Gloomy' – and a tough veteran of the first Scott expedition, Frank Wild, whose mother was said to be descended from Captain Cook.

Shackleton's first expedition was given the popular keynote of rugged, tough endeavour, but the main aim was the glittering prize of all exploration – first to the South Pole. In that opening decade of the century the bitter desolation at the end of the earth was publicly vaunted as a diamond-bright magnet for the brave British spirit – and so helped to enlist money from private and public purses alike.

Shackleton went back to the known doorway to the same route he had trudged with Scott. Although the dominant drive was to be first at the South Pole, he saw other challenges from which to win glory. There was the Antarctic's only active volcano, which no man had ever scaled. And in the west, in wasteland no man had seen, was the point of the intriguing South Magnetic Pole, that position to which the compass needles always turned, quite distinct from the geographical South Pole. Shackleton wanted complete conquest.

He sent the three Australians – middle-aged Professor David, Dr Douglas Mawson, and the Sydney doctor, Alistair MacKay to achieve both these objectives. Their initiation into the reality of Antarctic travel was staged on the slopes of the rearing volcano – Mount Erebus. They were given a support party to help them up the ascent with their heavy sledges, but the exploit was to demand courage and endurance.

Mt Erebus, the only known active volcano in the continent, soared 12,300 feet above the ice-plain, rising out of desolation, where, today, the US station of McMurdo (established for the International Geophysical

Year in 1957) stands – a modern township of more than a hundred buildings, with a summer population of some 800 men and women, and equipped with community services, a cinema, nuclear powerhouse, telephones, laboratories, and hospital.

It was daunting, back-breaking challenge. The laden sledge was hauled up steep frozen slopes, across rising fields of serrated ice, slippery wind-polished sheets, and areas of sastrugi and ice-falls with fissures and crevasses all around. Across fields of broken ice they often had to carry the sledge to gain ground. Upward they fought their way, with deepening cold and bone-cutting winds testing their stamina and resolve. After two days they reached 9,000 feet and were trapped for another two days in a blizzard. For the next forty-eight hours they lay shivering in their sleeping bags, wet, cold, hungry, unable to light the primus, knowing that to suck snow or ice at such low temperatures would crack the flesh of their lips and tongues, and cause intense pain to the alimentary tract. Thirst became an agony worse than hunger. The day after they could leave that camp, they climbed to the first of three known craters.

From there the uphill struggle was across ice-falls, through deep snow, an almost vertical ascent over ridged and sharp-edged frozen confusion to more than 12,000 feet. Ice crystals in the bitten air stung their faces and eyes, and caught in their nostrils. But, they found a geologist's paradise, an unknown crater with steaming fumeroles – open cracks – in the base of a vast saucer with spuming steam, which at once froze into delicate draperies of ice.

They found huge crystals of feldspar and rare rocks coated in yellow sulphur thrown from this boiling crater – three times deeper than Vesuvius and almost a mile, lip to lip. It was riven in its basin by a great fissure that ran down 400 feet to the bowl of lava fire. From there the spitting steam leaped a thousand feet into the air, along with fire-dust and the glowing rocks that were hurled high. Mawson was enchanted. The fury of the heated earth bursting into this frozen setting was the example for him of the wonderful contradictions and baffling complexity of nature.

They could not stay here long, despite the wonderment. Through broken cloud, base camp was a black dot in a wide rumpled cloth of white and lavender, the great ice shelf vanishing into distant haze – and to the west, behind the camp, sawing the sky towards the South Pole, was the endless range of the Transantarctic Mountains, a scene of alpine grandeur unmatched in the world… and beyond, and beyond, rolling westward for thousands of miles, was an unseen, untrodden, white land.

It was a glimpse all too brief for Mawson. He had to shield face and eyes from the biting frost and start the fight downward; yet, the awe-inspiring moment went with him, and brought a rare emotion that touches some southern travellers. In that ethereal setting his mind caught a shred of longing, a desire to walk that land beyond the great mountains, to explore its snow-wastes, coasts and uplands, to taste its timeless solitude.

It was a limpid stream of thought, he told himself, a gossamer that would vanish in the reality of the downward fight. But it never was banished; it lay on the edge of his mind to return again, and again, in the miles of sledging that lay ahead. It rested on the fringe of conscious thought to emerge and challenge his ambition until at last he was compelled to put the aspiration to the test of reason and positive action.

The story of Shackleton's attempt to win the South Pole on this 1909 journey is now a major classic in human exploration. With three companions – one of whom was Frank Wild – and supported in the early stages by four Siberian ponies, he slogged his way to within 112 miles of the South Pole and a plateau altitude of 11,600 feet. There the threat of starvation, incessant inroads by cold, wind, toil, and the constant enmity of the land drove him into unwilling retreat.

Like Scott, Shackleton brought back a clear warning of the implacable nature of the continent, but he also won some personal glory in the confrontation. As such, his journey overshadowed the greatest unsupported foot-slog ever made in the south – the march to the South Magnetic Pole.

–

The journey to the elusive southern axis of the Earth's magnetic field was made by the same three men who had conquered Erebus in 1808. David – then over fifty – Mawson and MacKay hauled a half-ton sledge across sea-ice into the north-west and then fought up a glacier and through the Prince Albert Mountains south of Cape Adare. Suffering from hunger and snow-blindness they made an incredible march of 1,260 miles, which included geology studies, and finally came as near as possible to the Magnetic South Pole as the simple equipment allowed. This was attained at a point registered as seventy-two degrees twenty-five minutes south latitude, and 152 degrees sixteen east longitude. When severe snow-blindness disabled David, Mawson took over leadership and for 500 miles led them back to meet the ship on the west coast of the Ross Sea – a feat that David praised all his days:

'Mawson was the soul of the march to the Magnetic Pole,' he wrote later. 'In him we had an Australian Nansen, a man of infinite resource, splendid spirit, marvellous physique and an indifference to frost and cold that was astonishing – all the attributes of a great explorer.'

The journey cut deeply into their physical condition and only by eating seal meat on reaching the coast were they saved from scurvy. Through thick snowdrift, bitter winds, over difficult terrain, they made long burdensome marches – and each of them swore they would be happy never to see this awful land again… yet Mawson's mind turned again and again to the dream on Erebus. Not for years was he to know that, while inflicting its ferocity on him, the Antarctic had already claimed his spirit and was calling him back before he'd departed. The challenges of the continent imparted sharpened intellect; mental and bodily effort were improved; but there was something else, something intangible in the total magnet of the southland. Years later Mawson was to write: 'We came to probe its mystery, to reduce this land to terms of science, but there is always the indefinable which holds aloof yet which rivets our souls…' The pull of the unknown on the soul of the explorer was irresistible, and Mawson was committed, in his own special way.

The Rejected Invitation

Homecoming, the brief relish of heroism and the repose of a workaday world quickly palled for Mawson. By 7 December, he was aboard the SS *Mongolia* en route for London. Days of rough seas across the Australian Bight, with winds blowing from the south across the open deck, and he was at once knitting the shreds of his Erebus dream into a pattern for action, listening to his 'faraway voices' and thinking of grander times trudging the snowy wastes. Fragments of thought and poetry filled his mind, of 'irresistible glamour of the south' and of the 'sullen solitudes' in vast 'God-like spaces.' He was entirely captured to his future.

In Perth, the Western Australian capital, he rushed ashore, and on 11 December 1909, knowing of Captain Scott's impending second expedition, sent him a brief cable saying he was London-bound and would call on him 'relative to Antarctic matters soon after arrival'.

One cold morning in mid-January 1910, Douglas Mawson left his rooms in Regent Street and rode in a hansom behind a steaming brown horse through the London mud and slush for his first meeting with Captain Scott. At the expedition offices in Victoria Street, knowing the reputation of Scott of propriety, and his authoritarian naval manner, Mawson was caught off guard by the warmth of greeting.

Scott came bustling out of his office door, face beaming up to the tall Australian, wringing his hand vigorously and ushering him into the 'sanctum', saying: 'My dear Mawson. How good to see you. I can't tell you how delighted, how glad I am that you are going to join us!'

Mawson was taken aback. In a flash he realised Scott's wrong interpretation of the Perth cable. He shook his head, and said there was misunderstanding. 'I'm afraid I'm not asking to join your expedition. I want to consult with you on how a special Australian party might be made an addition to your venture.'

Scott was now surprised. Not so much by the idea that a small party of Australians should join his expedition, as an appendage, but that Mawson had not come half-way round the world to join his expedition. With thousands waiting, here was a man who did not seek a place! This made the Australian hard to understand.

Mawson said: 'Have you given any thought to the exploration of the unknown land west of Cape Adare?'

This was not a welcomed member of Scott's mission; this was a man with ideas of his own. At once Scott fell back into his normal attitude of reserve and the warmth of that first moment of meeting was gone. Indeed, events already in train were to affect their relations, and cause complaints of lack of forthrightness, frustration, and dissatisfaction.

(In the previous April, the American, Admiral Robert E. Peary, along with Matthew Henson and four Eskimos, had reached the North Pole. Balked of his plans to reach the North Pole first, another explorer had secretively switched his objective to the other polar prize – the South Pole. His name was Roald Amundsen. How this development would impinge on their own lives and plans both Scott and Mawson were entirely unaware at this first meeting.)

Scott told Mawson the idea of exploring west of Cape Adare had not crossed his mind. Meanwhile he studied his visitor carefully, sizing Mawson in the flesh against his reputation with Shackleton. Tall, with clear blue eyes full of quiet confidence, he was at twenty-seven years of age a hardened, experienced sledger, and a scientist of resource and commitment. A man of stamina, a man for the South. Scott decided he wanted Mawson in his team. At once he invited him to lunch.

In a restaurant in St James, they talked and tried to urge each other to a point of view. Mawson argued the value of exploration of the great quadrant of untrodden land below Australia, for many reasons. 'That country is closer to southern Australia than Hobart is to the Western Australian city of Perth,' he said. 'Few people realise that. And while several missions have ranged through the Ross Sea, and have walked the ice shelf and McMurdo, and gone south to the polar plateau... not one human foot has been set on that land. I am compelled to go there!

'I am here to ask if you will take me, and three companions from Australia, and set us down at Cape Adare so we can explore the coast westward from there? If you will do that I will join your expedition with my group, as a corollary to the scientific work of your own expedition. Also, I want to make a proper job of the South Magnetic Pole and, as well, the geology might be of benefit to Australia.'

Scott hesitated, then conceded: 'I find the idea quite interesting, Mawson, and I will think about it and go into the ramifications.' Then, unwilling to leave his own want unsatisfied, he said: 'You feel you must go south again. I understand that, and I repeat my offer – join me in this mission. Look, I will put your name down this afternoon as a member of the staff. I will pay you not less than £800 for the two years. If you don't confirm acceptance within three weeks, I'll remove your name from the list...'

Mawson was unmoved. Scott then leaned across the table. 'I will go further,' he said. 'I will guarantee you will be one of my three companions in the final dash to the Pole!' First at the Pole! He was severely tempted. Scott saw that.

'Let's leave it for now, Mawson. I'll think over your idea and you think over my proposal. Come to my office tomorrow morning and we'll talk again.'

–

Next morning, Scott had the company of Dr E. A. Wilson. Mawson did not like that. Wilson remained silent for most of the interview, but Mawson felt his influence, as Scott's chief scientific officer, bearing on the outcome. The first blow came from Scott.

'We have our hands full with the McMurdo Sound plans, the march to the Pole, and for setting an eastern party around the coast of King Edward VII Land. That means, I'm sorry to tell you, we just couldn't possibly include a landing party on the north coast. We are pledged to this other work and cannot see any way to extend it.'

They gazed at each other across the desk. Wilson stood to one side. Mawson felt the atmosphere had changed to a rigid formality, that he was being manipulated. He stiffened his resolve.

'That's quite regrettable Captain Scott. I shall have to go to the north coast by other means, and I'm going – even if it is in a whaleboat.'

Here, Mawson believed, Scott took a defensive attitude at the prospect of an Australasian expedition in the Antarctic. He was surprised to hear Scott say: 'I must confess it has always been my intention to do what I could around the north coast – but could promise nothing. In fact, I think I may now set my mind on picking up some plums down there by simple reconnaissance by boat, when the ship returns after the winter.'

There was a pause in which Mawson looked both at Scott – then at Dr Wilson. He broke the silence, speaking directly to Scott: 'I must say, finally, that if I cannot be landed on the north coast at Cape Adare, I cannot join your expedition.' Scott showed his disappointment. Mawson added: 'As you will not consider landing me with three others at Cape Adare, I would only join you as chief scientist – and since Wilson has already been appointed to that position, I would not dream of seeking the post. Indeed, I'll say here and now that, hence-forth, I will not consider joining your enterprise.'

Scott was not done. He asked Mawson to dine with him and Mrs Scott the next evening, and Mawson agreed. When he left Scott's offices it was not yet midday. He went to talk to 'The Boss' – and at once Shackleton was bubbling with enthusiasm. 'I see a lot of worthwhile things to be done and I think you are right. I think you should go on your own bat – and I can get you some cash support.'

In the quiet of his room, staring into the night, Mawson knew he was at a cross-roads. Glamour surrounded the coming Scott expedition; men were to reach the bottom of the world, almost certainly. Was he to turn his back on a place in history – for his own dream? What was his alternative to 'pole hunting?' There was this unknown coastline at Australia's backdoor – and it had not long become apparent that the world had a sixth continent – the shape of which was still unknown! The first encampment on the shingle beach at Cape Adare had been only twelve years before – and there was so much to do, to find, that to reach a single point on the face of that untrodden land seemed meaningless. That highway to the south was now well-travelled. He told himself: 'I have seen the great mountain chain ranging south through Victoria Land; I have walked those ice mountains and seen rocky exposures and they could be of economic as well of scientific interest. No man has been there. They could hold as much mineral wealth as ranges elsewhere along this same great Andean chain that spans into America. It is a blank area on the map and while some will ask "what's the use of it all?", I must say Charles Darwin once looked at New South Wales and said it had no future – that, after Cook condemned the "doomed" land, men made fortunes at its edges. As well, this leads to conjecture that one day the frozen south might play a role in the arts and sciences of civilization. Every observation we make in that unknown will add to world knowledge. One day we shall realise its place in the universal plinth of things. I want to make fresh contributions to the human store – to reveal the structure, unravel the mysteries, study the geology, the weather patterns, the wildlife, the marine biology, the birds and fish and sea plants, to make a proper job of locating and measuring the South Magnetic Pole and to investigate this little-known phenomenon of the Southern Lights – the *aurora australis*… there is something enormously important in all that, I feel.' He fell asleep knowing then he would not march to the South Pole with Scott.

–

Dinner at Captain Scott's home was a carefully planned appeal to his ego and patriotism. Scott was especially complimentary; Mrs Scott showed her admiration of Mawson and joined the urging: 'I hear Con wants you to go south with him. Isn't that right, dear? It must seem a wonderful opportunity for you and one I should think you wouldn't have to think twice about. It is likely to be historic, you know.' At one point during dinner Scott waxed most appealing: 'Can't you see what it will mean, Mawson? We should be there together at the Pole to raise the Union Jack, for the sake of the Empire, for all that Britain has done up to now. Come with me! I want you to share that moment!'

Mawson was direct; his most blunt and realistic self. 'Raising a flag at some point on that plateau has no real meaning to me, though I must say I'll regret the loss of comradeship and the privilege of serving with you. But, Captain Scott, there are other things which call to me. They are more important than the notoriety of Pole-hunting. If I am appealed to nationally, I would have to say that it is not the polar plateau, but the vast quadrant south of Australia which is in my land of hope and glory.'

Neither man would meet the other. Still, Scott left the door ajar. 'I will leave your name on my stafflist until the day the ship leaves Hobart – in case you change your mind. If you are not going to, then I would be grateful if you or Professor David recommended an Australian geologist to go in your place.'

'But I tell you now,' said Mawson. 'I will not join your expedition! There is no hope of that.'

They shook hands standing on the doorstep of Scott's home in Buckingham Palace Road, each holding the other in regard and high respect, but neither yielding the principle.

For those last moments Scott's manner of authority vanished, replaced by the warmer quality of two explorers with respect for each other. They looked into each other's eyes, then Mawson walked away down the street. They were never to see each other again.

Mawson turned to assessing prospects of raising funds from wealthy Australians then in England. Early one morning, Shackleton burst into his office. Without preamble he declaimed: 'Mawson, me boy! I have decided to go to the coast west of Cape Adare. And you are to be my chief scientist!'

Mawson was astounded at the brusque announcement. Shackleton went on: 'It's like this. I can get the money and that will relieve you of the trouble of doing it all by yourself.'

Shackleton had realised the value in knowledge such a venture would bring, and was now eager to take it over. But what sort of expedition would it be? A few men in a single landing such as he had proposed to Scott? Mawson asked: 'What sort of money are you talking about?'

The deal was set when Shackleton said: 'I can get hands on £70,000 for the job.' It meant their own ship, a strong party, landings at several places along the unknown coast. It was a glowing prospect, worth yielding the leadership for.

They spent weeks together seeking funds and received promises of many thousands. Then Mawson heard Shackleton talking enthusiastically about a venture to Alaska, and that his wife was averse to his going to the Antarctic again. Slowly, doubt crept into the enterprise, to be briefly dispelled when Mawson noticed on 18 March, the London Daily Mail published details of the 'new

Shackleton expedition'. Shackleton bubbled excitedly: 'Whatever happens now, Mawson, that area is reserved for us by the unwritten law that explorers don't poach on one another's declared territory.' However, 'The Boss' was never to see the land west of Cape Adare. Full of schemes and ideas he went to America where Mawson eventually caught up with him in Missouri and drew up a contract on the night of 18 May 1910, stating that should Shackleton pull out he would still support Mawson's leadership. It did not at once clear up uncertainty and confusion; not until December 1910 did Mawson eventually learn that there would be no leadership by Shackleton as he was already planning for a transantarctic crossing.

By then, Scott had reached Australia, had asked Mawson for his detailed plans and had sailed south to the shock of finding Amundsen at the Bay of Whales, on the Ross barrier, preparing an assault by dog-teams on the South Pole. Freely sending his plans to Scott, Mawson had raised the point that he had refrained from seeking money in Australia until Scott's venture was on the ice. He took no action, he said, so that Scott could never regard him as a usurper of funds that might have gone to the British expedition. He added: 'It is with great sadness I write, for it seems to me that one great chance of a lifetime has slipped away. It is not the notoriety of reaching the Pole that I refer to – it is lost companionship of great spirits, that alone which makes life a harvest of joy raised upon a battlefield of suffering.'

He told Scott he planned to land a major party at Cape Adare and other parties along the coast towards distant

Gauss Berg. He wrote: 'I am conscious of loss incurred by dissociation from your leadership; for, believe me Captain Scott, I have recognized in you an ideal of generosity and humanness... My best wishes go with you for the safety of your expedition, for the alleviation of your hardships and for your final success.'

There was never any reply from Scott. He did not live to apologise for the landing of his second party at Cape Adare after finding Amundsen at the Bay of Whales. It was the news of this which, reaching Mawson when he was ill with influenza while fundraising in London, caused him to write to Scott's wife with some complaint: 'I *do* wish Captain Scott had been franker with me – instead of it all proceeding from my side!' He well knew the cause, for he said: 'I think it is quite unnecessary for explorers to act in the way that Amundsen has done.' But, even then, he could not accept that as full and valid excuse. 'Had Captain Scott truly desired to settle an eastern party at King Edward VII Land this year, it seems to me that the men he has had pluck enough to do it. The K.E. VII Land wants doing, and none but a strong expedition like that which Captain Scott has can do it. But, any old ship from Australia can land at Cape Adare... I'm glad for your sake that Captain Scott has much the best chance of getting to the Pole, though I think he will have a hard race.'

This was in April 1911, and though he was now certain he would lead his own team to the Antarctic, he did not know that Amundsen's and Scott's actions would drive

him far to the west of his objective and force him finally
to set down on terrain of which he would write...

We have found the kingdom of blizzards;

We have found an accursed land.

ONE

Driven Westward

The steel-clad nose of the little oak ship, *Aurora*, butted into the fringe brash of the southern pack-ice at four in the afternoon. On that day – 29 December 1911 – the Antarctic first embraced the Mawson expedition. They moved at once into a changed environment.

Millions of jostling, bobbing pieces of ice strewn across the surface deadened the motion of the sea and softened all sound so that a hush fell over the vessel. It was strange, eerie silence and stillness after weeks in the Great Southern Ocean where gales had battered them into gratefulness for a patch of blue sky, a glimpse of the sun, and relief from the hammering wind.

The ocean assault opened on the day they left Hobart – 2 December – to make a nightmare of the 900 miles journey to Macquarie Island where Mawson was to land an exploratory party of six who would operate the radio relay station. On that first day a mountainous sea staggered the *Aurora* – a thirty-five-year-old Dundee-built veteran of the Newfoundland sealing fleet. It carried away half of her bridge, soaked the Greenland huskies tethered on the deck, and damaged other superstructure. The seas and strong winds made the landing at Macquarie Island both dangerous and arduous, and delayed the work of hauling radio masts and equipment 300 feet to a sheer cliff-top. For the next thousand miles south men were thrown from their bunks into inches of cold, inboard sea water, and havoc was caused by the 380-ton vessel rolling, pitching, and juddering as her coal-fed ninety-eight horse-power engine strained to keep her head to the wind. For days, upheaval put the small galley out of action. And to add to the miseries of cold food, there was also thirst when the sea burst into their fresh water supply. Nobody washed, or changed clothes and drinking water was rationed; and with the strain of sleeplessness on them they longed – Mawson and all his twenty-five recruits – to reach the ice.

Now they entered this coldly muted setting, with the weeks of southern summer fast flying away. Pushing south through the ice in these uncharted waters, Mawson clung to a hope of winning an open passage directly down along meridian 156 east. Southward on that line might bring him to a hoped-for landing site, a hundred miles or so west of Cape Adare, west also of that vast glacier-laced mountain chain he had seen from Erebus in 1908. If a passage was found direct south, his far eastern party would explore that land beyond the great Transantarctic Range, which runs as a huge ice-coated barrier to the high plateau of the South Pole. West of those mountains was the terrain that had called him since he was with Shackleton.

Mawson stood on the battered bridge, peering forward, hope dwindling as the brash grew larger and noisier under *Aurora*'s prow. The frozen pieces went scraping and crunching along the oak planking, swirling astern with the four-knot speed. Then, the clinging fog rose and draped itself like a diaphanous skirt over the ship, leaving filaments trailing from the masts, the yards, and rigging. Suppressing sound and sight, it added to the peril of a quiet, ice-strewn sea that could quickly freeze and trap the ship, and the expanding ice would grind the *Aurora* to matchwood.

Still, they pushed south. Mawson stayed on the bridge as the hours went by, watching Captain John King Davis, his second-in-command and master of the vessel. The captain knew this ship — he'd bought her in St John's, and nursed her at six knots across the world. He was tense, wary, peering constantly ahead, his face — more than ever like a cadaver — thin mouth tight — wrapped in a grey balaclava. It was a haughty face that rarely broke into smiles. Davis fully deserved the name the men had given him: 'Gloomy'. But they were all lucky to have him aboard. One of the most reliable and experienced Antarctic navigators, sharp, alert, he was ever in undisputed command of his ship. Mawson listened to the captain's voice crackling into that stilled air. 'Steady! Steady as she goes.' — then, suddenly snapping: 'Hard a-port!' The rudder chains lifting and clanging back in their steel channels set the frightened dogs howling in the deck coops, and as the ship swung away so they glided past their first massive iceberg, an Antarctic flat-top that loomed through the fog like the phantom of a great white building.

Mawson's admiration for Davis grew that first day among the ice. He had earlier shown superb skill and judgement in the ocean; Mawson believed he had kept the *Aurora* afloat when others would have been capsized and shipwrecked. Here, his skill was even more pronounced, his calm and his concentration unbroken. The red-haired Irish captain did not – could not – relax in these waters. Like Mawson, he knew there was surface cargo that made the *Aurora* a potential firebomb. On the poop-deck, lashed down in five-gallon cans, was 6,000 gallons of high octane benzine, fuel for the Vicker's REP monoplane – the first aircraft to go to the Antarctic. There were other inflammables aboard – fuels, oils, kerosene, priming alcohol, dynamite, cartridges – all menaces to safety if ice sliced into the cargo of benzine, if a spark from the smoking stack fell into spilled fuel. And an underwater frozen knife-edge protruding from a passing berg could rip away the vessel's bottom.[2]

Mawson and Davis stood watchful on the bridge knowing they were thrusting into these waters at the worst time in the southern summer, when the prevailing south-easterlies across the wide Ross Sea broke up the ice westward from McMurdo and from around the Balleny Islands and streamed it across their southward path. They could not wait until later; there was so much to do, so far for the ship to travel before the autumn freeze set in.

---

[2]The ability of ice to maim a ship was to be demonstrated in the Atlantic two months later with the sinking of the *Titanic*, the biggest liner afloat.

The plan, at that time, was to settle at least three groups of men at stages along the 1,800-mile coastline, between the 156 east meridian and the land sighted by the Russian, Drygalski, near Gauss Berg, around 95 east meridian. To delay this southern sortie to later in the summer, in the hope of clearer water, free of ice, would endanger the prospects of the western party and perhaps imperil the ship itself. To have come earlier would have meant confronting a solid barrier of ice. If the floes and growlers parted to allow them through to land then the disruption caused by Scott's changed program would be minimised. This day, 29 December, thus became a key period in the Australasian Expedition of 1911. If the ice had then allowed them to pass, the whole series of events, the tragedy, the incomparable fight for survival, the strike south to the Magnetic Pole, and the exploration of new lands west, would all have been changed.

Douglas Mawson thought of the last two years: memories of bluff, hearty Shackleton, of Captain Scott and his clipped naval manner, of the frantic, hectic, worrisome days of fundraising, the organisation, the rebuffs from autocratic people in London who believed he was a fly-by-night aiming to steal Scott's thunder by a secret dash to the Pole. That was all past and now he was among the ice, his expedition mounted and almost paid for, equipped to land twenty-six men on the mainland and enough fuel to run the *Aurora* so she could come south again next summer and take them all home. He had gathered a fine team of men, had won support from scientific groups, including the Royal Geographical Society and the Australian Government; he had a tight ship, a fine captain, and among the young men from Australian and New Zealand universities without polar experience he had the stiffening of the veteran Frank Wild – who had sledged furthest south with Shackleton – and the Swiss-born world ski-champion Dr Xavier Mertz to support his own expertise. But, he well knew that the awards, the degrees, the goodfellowship of the common-room could fall away in the confrontation of the ice if the basic steel in the character was missing. At the end of a year, they would all stand revealed.

There was a stirring of wind. The fog lifted and the scene stood clear ahead. The vista southward set back their hopes; Davis eased the engines to a slow drift forward. They had been four hours in the ice, twenty miles or so from meeting the first brash. Now they were beset in a sea of flat-faced hunks, floating alabaster, milky-white slabs laced with shades of lilac and mauve, and – where some had been overturned – daubed brilliant yellow and ochre with the algae diatoms that live in southern waters. There were huge icebergs; one running clear across their path was more than a mile long with frozen turrets 200 feet above water, its supernal sides of blue-green ice laced with honeycomb caves and with great caverns at sea level into which the water boomed and splashed. Much further south they could see where the ice was hummocked, great pieces prised one atop the other for lack of room. The *Aurora* was denied a chance of a southern passage.

And over all this came the penetrating hiss, the song of the pack ice; millions of frozen faces rubbing together and oscillating in the subdued swell... an insistent, menacing sound, which Mawson thought was like the wind soughing through the tops of parched eucalyptus. It was the song of defeat. The ship quivered with the blows of ice slabs and the single propeller was in danger. It was time to turn back to the north.

It was late in the evening; the imminence of the midnight sun made the misty air translucent and blown spindrift on Davis's woollen balaclava had frozen into glinting sequins. The captain called the order to go about. They turned back, disappointed as they would be, again and again, in the coming week.

Day after day they were driven westward, always probing southward to seek a way through the ice barrier; searching the sky each day for a break in the continuous 'ice blink', the pale ochre glare reflecting and filling the sky from the endless frozen sheet over the sea. If there was open water ahead it would give a different reflection, it would be a 'water sky.' However, there was only the pale glare ahead and the sound of the rubbing faces of the floes to remind Mawson of the all-powerful force lying in wait to freeze suddenly and clutch the vessel into a final embrace. The ice held only one blessing; parties could land on the bigger floes so that blocks could be cut and hauled aboard for melting into clear, fresh water, to ease the rationing.

As the days went by, they saw more and more wildlife. Whales blew in plenty, and a few awkward sea leopards and crab-eater seals lolled on larger floes. There were flocks of birds – clouds of beautiful snow petrels, the delicate Wilson petrel among them. They saw great hunks of ice with rocky debris embedded and hoped that these were signs of nearby land. But, late on 3 January, the pack ice barrier again swung north, forcing the ship into higher latitudes… mile after mile they were compelled to sail away from the hoped-for landing.

They were some 800 sea miles west of Cape Adare, and still with no sign of reaching the mainland coast. Mawson was growing anxious. Macquarie Island, with its vital relay radio link with the mainland, was more than a thousand miles away; they would be stretched to their limit of power to transmit across a distance much further than that. The six men landed there, under the command of Ainsworth, were well-provisioned, but, if they were beyond his radio reach, he would be without the communication links he'd planned through the mainland station, on the Bluff at Hobart, to inform supporters and the public of the progress of his mission. They were his men, and he was responsible for their safety and well-being. There was yet another factor of growing concern, the South Magnetic Pole work. Joint studies had been arranged with other people; his magnetician, Eric Webb, needed to operate a program as near to the South Magnetic Pole as possible. Each mile sailed west now made that more difficult and added sledging miles to the journey that would have to be made into the unknown hinterland.

How many more days could they spend probing through the pack in this fruitless hunt? How many weeks were left to sail the ship westward to land Wild and his party? Not just to get men ashore before the sea froze over and captured the vessel; to get them ashore in time to build the huts that would keep them alive through the winter.

He held a conference with Captain Davis. They pored over the only available sketchy maps.

Mawson said: 'We have to face it. Our ambitions must be limited. I saw no reason to think we couldn't reach a rocky or gravel beach like that at Cape Adare, and now we're hundreds of miles west and still blocked from the mainland.'

Davis bent over the best available chart, fingers tracing the journeys of the Wilkes' US naval squadron and d'Urville's route the previous century. There were soundings listed, and few lines of reported sightings of land, but these were not reliable; the *Aurora* had already sailed over land marked on these charts. Davis now stubbed his finger at a marked headland... on the 140 east meridian, a line intersecting a pencilled curve: 'Here's where d'Urville reported seeing a cape in what he named Adelie Land. If it *was* rock it would still be there – if the ice hasn't blocked it in. And, it's still some 200 miles west.'

Mawson brooded. Two hundred miles along the ice-bound coast, prodding and probing, and a top speed at the most of six knots... it gave him a few more days. The decision was forced on him. 'We'll go no further than there without taking desperate measures,' he told Davis. 'We must get ashore – somehow. Then you must hurry west with a strengthened party of eight under Wild. We can only have two stations now – at the best.' Then the expedition leader and his second-in-command went to

45

their beds, disappointed, feeling disaster stared them in the face.

-

At five next morning the prospect suddenly changed. The watch woke Davis and Mawson, excited: the air was clearing and there seemed to be land ice close by. Sure enough, on the port side, 400 yards away, an ice barrier rose solid and sheer with only fragments of pack in their path. In each direction the ice wall disappeared into the mist, a vast sheet of massive ice. They followed it north-west and after two hours steaming turned its corner to run south.

As the scene unfolded before him, Mawson happily exclaimed: 'It's an Eldorado!' Wind had cleared the pack from the lee of the ice wall; they broke through into clear water, steering close to the rising face, wondering at the formations and colour of the mighty edifice, disturbing flocks of birds nestling in its sides. Now the sky was open, the water was lit by bright, low sunshine; a glorious day for discovery! There came flocks of wheeling Cape pigeons, Antarctic birds of all types; and seals leaping from the ice floes. Soundings gave a depth of 280 fathoms, 1,680 feet of water to a bottom they guessed must be the Antarctic continental shelf. The ice-wall, rearing high, turned sharply south-east – but, a strong wind coming from the wrong quarter would have driven them onto the craggy prominences; once more the Antarctic snatched away hope. The white face ran away east into foggy distance. Davis turned the ship back to the lee of a large iceberg to wait on a more favourable wind, but in the evening it blew a strong gale and all night they steamed up and down in the lee of the berg with flurries of snow whirling over them and coating the ship white.

Not until the afternoon of 5 January did the weather abate. Then they entered days of discovery. Their great ice wall swerved away into the east and proved to be a colossal, floating, frozen island, easily the most immense berg reported, more than forty miles long, and which had not long before been part of the continental ice.

The *Aurora* sailed south-west to confront another immense ice wall. This time Mawson, Wild and Davis recognised a familiar pattern. This was a glacier tongue, thrust out along the sea-bed by remorseless pressure from the mainland. The bottom was grinding along in mud – 395 fathoms deep under water; ice, 2,370 feet thick below the surface! Above the water, it rose to heights reaching 200 feet. A vast glacier. Seaward it stretched out sixty miles and there, where it floated, its tongue broke off into huge ice islands like the one they had encountered the previous day.

Mawson exulted in the find. 'Within gunshot is the greatest glacier tongue known to the world,' he wrote. 'No human eyes have scanned it before ours… the feeling is magical.' The young men of his party rushed on deck, some only half-clad, and, in the biting glacial wind, they danced with excitement. Mawson understood, and wrote of this thrill of discovery: '… the quickening of the pulse, the awakening of the mind, the tension of every fibre – this is joy!'

It was more than joy for him; it was relief. The glacier was obviously fed by continental ice, and that meant a chance of a landing and banishment of his looming worry – of being defeated by the pack and not effecting any landings, perhaps even returning home with nothing achieved.

Suddenly he felt the cold long hours in the open, gazing at the glacier, had made him tired. As well he felt a need to express his relief; and, in the loneliness of command, he went to the engine room, to get warm – and to write to the girl waiting for him in Adelaide... to Paquita Delprat[3], the dark-haired beauty he had asked to marry him some eleven months before... 'This morning I wished I had never spoken to you that quiet evening on the verandah at Brighton...' Now the threat of disaster and defeat were allayed, they were *near* the mainland. 'Already we have made important discoveries, and – Oh, my dear – I am beginning to live again, after days of impending evil, and disaster ... we now have a lead to the south!'

He was too optimistic; again the nose of the *Aurora* was turned westward. The glacier tongue joined with the ice sheet, and at this junction the frigid cliff rose sheer out of waters where jagged ice ridges ran side by side with rocky reefs. They could not take *Aurora* inshore. Where these steep cliff faces fell away to bluffs or coastal valleys the ice was shattered, broken into a confusion that made a landing impossible.

---

[3]The daughter of G. D. Delprat, Dutch-born mining engineer who was then manager for the Broken Hill Pty. Coy. Ltd.

On the eighth day of the new year, they came at last to the sweep of open water that he was to name Commonwealth Bay. Scooped wide out of the ice, fifty miles across, it was bounded eastward by the wall of the glacier and on the west by a distant cape that might have been that which d'Urville had named Cape Decouverte, the only part the French explorer had seen of the land he named for his wife. Here, too, the land itself was smothered by the great icecap and there was no sign of rock.

Through this wide bay they wended a slow passage, scanning the way ahead with binoculars, growing more depressed with each passing hour. Fewer than forty miles away now was his limitation – the 140 east meridian, the eastern-most point of the territory claimed for France. Mawson didn't care who claimed the land. So long as he could get ashore, they could lump all this new land into the Adelie territory.

The day crawled slowly past. He went to the cabin with Davis to drink hot coffee and talk. Then – Wild was in the doorway, his manner apologetic, voice calm, but his eyes gleaming.

'Sorry to interrupt,' softly, with a hint of a smile. 'I think perhaps you'd better come. I've just seen a rocky site through the glasses. You might want to take her further in for a closer look.'

Davis edged the *Aurora* to within a mile of the rocky strip. A boat was lowered and Mawson and Wild went ashore, making their way through a surprising cluster of little islands, some draped with blown sea-ice that dressed them like fancy cakes, others that teemed with penguins and seals; flocks of birds wheeled over them, skua gulls, prions, Cape pigeons, and a type of Antarctic petrel never seen before.

Beyond these islands – named the Mackellar Islands for an expedition donor – they found a miniature harbour with a shallow bottom thickly strewn with luxuriant seaweed. All at once the sun peered over the inland and lit the scene; before them was a ramp of ice, a frozen jetty.

Mawson was first ashore, Frank Wild followed. They stood silent together, two veterans knowing theirs were the first human eyes to gaze at this long-lost land, an apron of terrain, a mile and a half wide, littered with jumbled rocks and the sheet ice that bound them together.

There were a few flat rocky stretches where the huts would go if they could blast the footings down into the tough gneiss rock that predominated. All else was ice-bound. On either side the terrain moved away to join soaring ice cliffs, and the way was fretted and fissured with crevasses.

Behind the little arena, rolling and rising, Mawson saw an immense mantle of solid ice, lifting into haze, reaching several thousand feet – on and on it went, beyond his vision, out into far unseen distance. The unknown main-land of Antarctica would be at their back door. 'My land of hope and glory,' he had told Captain Scott. Here it was! The last chance. Still, he hesitated, looking at this desolation.

It was forbidding. Yet, it was a site – the first offered in a thousand miles, and even if the shallow water, reefs and the myriad islets made landing operations difficult, there were spots for the huts. There was a tough road up the ice-slope, yet, it was a way into the inland; there was also a possibility of sledging along the coast over the frozen sea. There was a small headland here for the radio masts. His decision was made. The headland would be named Cape Denison, after a contributor in Sydney, and Cape Denison on Commonwealth Bay would be the site for his main base.

He looked again over the soaring coastal cliffs, back up the ice to the inland haze covering this mysterious hinterland. This was to be home to eighteen men and the dog teams… and no more time to ponder, now. He noticed the wind was rising, so they rowed back to the ship with their news, climbing aboard soon after eight in the evening. They were all tired, but hours were precious now; there could be little rest.

–

The landing started immediately a sheltered anchorage was found for the *Aurora*. Their small motor-boat puttered back and forth carrying cargoes of food, eggs and sheep carcases, tents and lumber for the huts, tinned beef, gear for sledging, tools, a small furnace… until the wind intervened.

It came charging down the ice slopes with a sample of what was to come, whipping their little harbour into a foaming upheaval. The motor-boat was rescued from ruin by strenuous action on the lee side of the ship, and the full strain of the wind on the *Aurora's* anchor cable was fierce enough to straighten a hook of steel with a two-inch diameter.

Recurring gales from the ice plateau converted the next ten days of unloading into a purgatory of straining muscles in two heaving, tossing, small boats. When the gales threatened ruin to their whaleboat and motor-boat – they were hoisted aboard and the men slid into their bunks, to be called an hour or two later when the tempo fell. For ten days without proper sleep, they hauled, rowed, lifted. They were fighting for time – a priceless factor to Mawson, Davis, and to Wild and his seven men.

What they achieved was in defiance of the cruel land. By 19 January, their two piles of timber for two huts, the aircraft, the three sixty-foot radio masts, transformers, electric motors, batteries, furnaces, stoves, equipment for sledging, the struggling dogs, scientific instruments, food for eighteen men for two years, bedding, blankets, tools, nails, fuel – including twenty-five tons of coal bricks all in bags, oil drums, kerosene – and books and papers and personal gear... were all on the rock apron.

That afternoon the men ashore were all called by Mawson aboard the *Aurora* for a farewell gathering. He used the occasion to discuss with Wild and Davis the ideas he held about western operations. It was a hope, he said, that Wild could get ashore about 500 miles to the west and the two parties could work toward each other. (In the event, the ice again dictated, and Wild and Davis travelled nearly 1,400 miles before they could land – straight up a 150-foot wall of ice – onto the great floating, frozen platform that Wild was to name the Shackleton Ice Shelf.)

Mawson's men and Wild's group sat in the wardroom for their last meal together, which grew into a party. Frank Wild somehow managed to dress up as a cavalier, with songs, jokes and recitations. Mawson produced a donation of special wine – it had been carried on the *Challenger* cruise of the Antarctic by a British party in 1872. He called them to silence.

'We can this day claim the discovery of new lands. We have walked where no men have trod before and we will go on to discover more new land.' Now he asked them all to stand and drink together before parting. He spoke the toast, and said: 'We drink to the health and safety of all Antarctic travellers – whoever they are, where-ever they may be.'

It was the evening of 19 January 1912 and some 1,500 miles south of Mawson's base, Captain Scott and his four companions – Wilson, Bowers, Evans and Oates – disappointed on the forlorn plateau at the South Pole, had started the fatal struggle northwards.

On that same evening, as his men settled into tents or bedded down in the Benzine Hut, Mawson walked the slope of ice to watch the *Aurora* dissolve into cold mist beyond the headland, leaving only a flattening trail of smoke to mark her passing. If all went well with Wild she would be back in Hobart before March and would come south again in a year to take them all home. He walked further toward the lifting ice and scanned the scene with a geologist's eye.

Beyond, on the vast austere plateau, there were ominous signs. High up, perhaps more than 2,000 feet, he could see horizontal shutes of snow, like streams of white candy-floss, blowing out into space, carried out of sight high over Commonwealth Bay. And the great, blank-faced, slippery canopy of ice was barren wilderness – frozen wasteland. This was among the worst known in the Antarctic – in the world. It was a savage overwhelming by ice far more extensive than any seen in the Ross Shelf area, or McMurdo, or reported from elsewhere.

Inland, it rose out of sight; on either hand the ice-fettered coast spoke of a hostile climate, of severe glaciation.

Mawson told himself: this is a living Ice Age! This is how North America and Europe looked 50,000 years ago when frozen sheets clamped the land and the glaciers ground out great valleys and ravines. Adelie Land was gripped in such an ice age. It posed difficulties – and it presented mystery. It meant inevitably that the summer was short. The weeks for sledge journeys would be limited to a shorter period than was possible in the Ross Shelf region. And it was puzzling that for all this terrible glaciation it could not be much colder here than it was elsewhere in the Antarctic! Did it mean that snowfall was heavier in Adelie Land? And if so – why should that be? He climbed into his sleeping bag for his first night in that *terra incognita*, wondering at the reason for the vast areas of immensely thick ice-sheets. Before many weeks had gone the continent would provide the answer.

TWO

The Kingdom of Blizzards

After their first night, in reindeer sleeping-bags stretched on the frozen ground, Mawson's main base party woke with aching muscles and chilled joints to face the slavery of carving a sanctuary in a glacial wilderness. Yet, the first morning ashore treated them to a scene of desolate magnificence. A rare, low sun glinted from the curving ice plateau and lit the bay to shimmering blue in which milky-white icebergs with shadows of deep sienna bobbed and curtsied in a gentle swell. Mawson found the light intense, the colours striking; fleecy clouds were painted in luminous orange in a vault of sky that deepened on the horizon to a surprising emerald green. The rocks, among brilliant white snow and ice, were moulded from black velvet and housed nestling skua gulls. In the kelp-filled shore waters seals sported, and here and there groups of earnest penguins bustled like miniature nuns on a mission, pausing only to stare at the curious humans who had invaded their world.

Eighteen men squatted in the rough hut, built of cans of benzine with boards and a tarpaulin for a roof. They cooked porridge on the primus stoves, and ate it with hard tack biscuits and jam, washed down with mugs of hot tea.

When he had eaten, Mawson stood among them and said:

> 'We must build our home – and we must
> build it quickly and well, or we are dead men.
> We are far from any trades hall and there will
> be no union nonsense here. Working hours,
> with breaks for meals and bad weather, will
> be from seven in the morning till 11 at night.
> After that you can sleep.'

In the days and nights to come they understood his argument.

He sectioned them into work gangs and at once the labour of clearing a site for their huts was started; rocks were man-handled and levered clear, the surface bashed into a fairly levelled plane.

They had to sink yard-deep foundations for the bolts of timber on which the floor base would be screwed. Pickaxes bit into the tough gneiss rock – rock from the earliest ages of the planet. It resisted at the cost of bruised wrists, jarred fingers, blistered palms. Mawson produced dynamite, but it was frozen and had to be carried in their shirts as they worked before it could be used. When the long drills bit holes deep enough to blow a cavity they could find no soft soil or clay in this benighted place; they packed their charges and detonators with penguin and skua gull droppings scraped from rocks.

The body-thawed dynamite and the guano worked well. At supper-time on the second night the wood footings were sunk into cavities, filled with rubble bonded with ice, from water poured into them that froze immediately in the biting wind slicing down from the glacial slopes. On their third day the floor plates were laid, and their fourth day was spent carting boxes of loose rock; more than fifty tons they carried, to pin down the floor plates and hold them bolted to the foundation stumps. For the first time that day they used the dogs, in two teams, training them to pull loaded sledges – as they would do at the end of winter.

Mawson watched his men carefully, unobtrusively. They suffered split fingers, frostbite, aching backs and sore muscles – but they enjoyed the days. They were young, mainly from Australia and New Zealand, soon inured to hard labour, cementing companionship with toil, sleeping shoulder to shoulder on the cold rock, quickly becoming hardened, tough campaigners together. Each man had been carefully selected. Yet, Mawson knew well – and had told them all – that the final, true test was the trial on the ice. He saw those who straightened their backs more often, or wrung their hands while others laboured; he noted that those who were at the heart of the work, who showed initiative and zeal, were always among the volunteers. He saw the processes he had experienced with Shackleton; capacity for sudden, deep sleep after tough, physical exertion, ravening appetites growing into longing for rich, sweet, and fatty foods to satisfy the demands for energy, to meet the heightened needs for body heat.

They were growing to whipcord toughness and they would need to build muscle and hardness for the trials to come. Their fine edge of physical excellence sharpened wit and intelligence; meals were marked by brisk repartee and boisterous banter, sometimes to excess. Every man won a shortened version of his name or a special label. Mawson was named Number One Dux – or 'D.I.'; Bickerton, the British aero-engineer, became Bick, Eric Webb was 'Azi' for his work with the azimuth; and the fresh-faced Lieutenant Belgrave Ninnis, became 'Cherub'. His companion dog-warden, Dr Xavier Mertz, was reduced to 'X'. Mawson held deep respect for Mertz, and being the only European in the Australasian Expedition, he was also due for some sympathy. Mertz tried to learn English; he was fed with outrageous swear-words and epithets as though they were part of normal vocabulary. At meals when he came to use them the whole company would burst into gusts of laughter; Ninnis would patiently explain, with a little German and a little French and Mertz would glare at his tormentors: 'A joke ist a joke, ja. But too much is enough!' And the laughter would break out again. Yet, they adored him, respected his ski-ing expertise, and liked his wide smile, white, strong teeth flashing under the trim black moustache. They also liked his cooking, though he always tried to avoid eating meat, being a near-vegetarian. As well, Mertz was always at the heart of the tough work and, with Ninnis, worked ceaselessly to care for the dogs.

Six days of toil had the walls erected and the roof frame set. Then the gale winds blasted down from the inland.

The roof timbers had warped with sea-water soakings on the journey south in the *Aurora*; cutting, planing, trimming in the Antarctic gale was a trial for them all. Vital pieces of wood were always missing, some lost under cover of the sudden snow squalls.

Two weeks after landing they had the roof in place and the wooden bunks screwed to the walls. Now came the time to fit the centrepiece of their winter existence, the stove. Their shield against the fearful low temperatures to come, vital for cooking and melting ice to drinking water – it was incomplete. The box of fittings, fire bars, flue damper lid, operating handles, door, could not be located. Then someone remembered seeing something fall from the tossing motor-boat during the landing and they scoured the shore near the ice ramp – and tangled in streaming kelp under seven feet of water was a wooden box.

Grouped on the edge of the near-freezing bay water, the men debated how they could secure a hook into the box and pull it ashore. Mawson, without a word, stripped off his clothes and plunged into the icy water. Breathless with cold, coming up for great gulps of air, he tore away the fronds of kelp gripping the box and after three mighty attempts, lifted it to his chest level and staggered onto the ice. Magnetician Eric Webb took the box from his arms, with a sudden memory of his first meeting with Mawson in the blazing sun of the South Australian outback, when his long, muscular arms were effortlessly swinging heavy boxes of uranium samples into a rail waggon.

They hurried Mawson back to the hut; Mertz lit a primus and heated cocoa. Mawson sipped the hot drink and grinned at his stunned, quiet men: 'It was not a bit exhilarating,' he said. He felt their discomposure, but he did not mind the discomfort, for, by his deliberate act of leadership they would know that Mawson would never ask them to do what he would not do himself.

They eased the situation by turning to the box, prising off the soaking wet lid... and the hut was filled with uproarious laughter – at Mawson's expense. The box contained four dozen tins of Australian jam.

The missing parts of the stove were found later, under a snow-covered pile of goods, and by early evening the coal was burning bright in the centre of the living quarters, the hut was filled with the delicious aroma of roasting mutton, and the men stowed their personal possessions – books, family photographs, diaries – along the shelves. The meal that night was a festive event – the first in the home they'd toiled to erect. Yet, some of them felt it had a special quality and atmosphere generated by Mawson's act of leadership, that he had bonded the men into firmer companionship. They all turned in early – except the nightwatch that was to be maintained throughout the mission, to keep the stove burning, and to monitor instruments – and every man revelled in the luxury of blankets in his wooden bunk, his own special place in the cold world.

The plateau winds increased their menace during February. They brought grave misgiving to Mawson's mind. One day came up clear and fine and, with sun and only a stiff breeze, it was like a gift from heaven; he thought of the many halcyon days like this he'd known at McMurdo Sound. He did not tell his men of the difference, and allowed them to accept the Adelie Land weather as normal for the Antarctic. And he did not allow the winds to delay the work. The second smaller hut – two huts because two parties had been combined – was erected and married to the larger hut. The smaller hut was a workshop and store, and served as a vestibule against heat loss from the main living quarters.

The snow-drift was soon higher than the five-feet high verandahs, making snow-walls for extra store rooms and a sheltered home for the husky dogs. As the weeks flew by and the snow grew deeper, the men tunnelled through it and their winter home became a labyrinth with tunnels to reach the outside world – there were tunnels for waste disposal, tunnels for storing gear, meat, and foods.

One day in mid-February, Eric Webb sounded a warning; a monstrous sea elephant – rare on the Antarctic mainland – had lumbered ashore and was threatening the dogs. When they reached the spot, Mawson saw the enormous, ugly beast standing over the dog Ninnis had named Johnson – after the famous prizefighter. Johnson was true to his name – he stood up to the huge animal – but then the sea monster saw the men as a tastier and more satisfying meal than a mere dog. He lolloped toward them; the biologist and physician, Dr J. G. Hunter, shot the animal through the neck, prizing the skin and the skeleton as a gift for a museum. The sea elephant measured seventeen feet six inches from nose to tail and twelve feet round the girth. It weighed many tons, and it was a prize in another sense. Mertz and Ninnis flensed the great carcase before it froze. Blocks of blubber for cooking and for stoking the stove fire to life in winter, and more than one ton of meat for the dogs were cut and carried into the deepfreeze of the snow tunnels.

As weather allowed, they built a hut of rocks and packing-case boards (which the gales blew down once) a quarter of a mile from the hut to house Eric Webb's magnetic equipment. Constant and regular readings would be adjuncts to the planned journey in the coming spring to locate the South Magnetic Pole. It was placed this distance away to avoid any interference from electrical gear or metal at the hut; it was a short distance in fair weather, but, in winter, it was to prove an arduous and, at times, a perilous journey. A hangar was also erected for the REP monoplane, using one wall of the hut. This measured thirty-five feet in length, to shield the fuselage, and was soon encased in snow. Mawson renamed the aircraft the 'airtractor'. Its wings damaged in transit it could not fly in Antarctica, and so was to be used on skis as a tractor to tow sledges onto the plateau.

Since this was a scientific expedition, there were outdoor shelters to be built for the anemometer, for all the measuring devices dealing with direction, the strength, and temperature of the winds; there were to be studies of the sea tides, the southern lights (or *aurora australis*) and their link with magnetic storms – and studies of wildlife, when this was possible.

They were given examples of the savagery of survival on the fringe of this cruel continent. Big Antarctic petrels, gluttonous skua gulls, hovered like flights of vultures over the carcases of dead seals or penguins, fighting each other off as they ripped the prone bodies to pieces with their sharp beaks. Mawson wrote into his diary: 'It is a diabolical sight to see the flights of birds tearing out the viscera of a seal, dancing all the while with wings outstretched, and screeching angrily at each other.' The same birds were corsairs of the air, raiding the nests of the inoffensive penguins, carrying off their young and their unhatched eggs. In the waters of the bay were packs of killer whales, fast and fearsome wolves of the sea, which slaughtered any careless seal or young sea leopard. Very soon, it was a hard world in another sense; the ice spread across the harbour at Cape Denison and rapidly wiped away all traces of wildlife – except the yellow stain of the guano. The last penguins went scuffling across the frozen sea, the birds flew north to island roosts, and the marine animals submerged to disappear to their winter haunts. They left the white wilderness and the desolation to the men and their dogs.

Mawson used bad-weather days for training his men for the exploration in the coming spring; familiarity with sledging equipment was gained by trial assembly and packing of a sledge in the hut. The trail-blazers had to be able to pack and unpack their loads, manage their portable life supports by feel and instinct. They had to prepare to make camp in the dark, in the white-out of thick drift, with faces virtually smothered and fingers bitten with frost. Innovations, changes in gear and design, came out of the patient assimilation. The eleven-feet long sledges of Australian hardwood and lightweight platforms built in of bamboo – fore and aft – and a spar and short mast were stepped to carry the groundsheet adapted as a sail when wind was favourable. The front bamboo platform housed the cooker assembly – one cooking tin inside another containing the primus – strapped securely to the frame; the rear platform took the fuel, gallon cans of kerosene, and bottles of the alcohol priming spirit. Mawson had settled on alcohol as more reliable, lighter, and otherwise more useful than the normal methylated spirits. It was a decision that helped to save life. In the well of the sledge went the packs of tent, with poles, clothing, ropes and harness, for both men and dogs, the tools vital to travel – ice-axes, pick-axes, spades, hammers, files, ice nails, and repair kits for tents and clothing, and medical box. There was a corner for spare footwear, gloves, for alpine rope and a small rifle; a spot for the theodolite box for surveying, for the scientific aids to read altitude, for compasses, and books, for the carefully balanced and apportioned food bags. Mounted behind the sledge was a wheel, like a cycle wheel, which measured one yard each revolution and which, geared to an attached sledging meter, gave a reading on the distance travelled. This was advice as vital to Antarctic travel as knowing the time and the direction. On occasions, the sledge meter was more crucial than the compasses, which proved unreliable so close to the exact South Magnetic Pole where the needle, instead of pointing horizontally in a general direction, is being dragged *downwards* by the force at the axis of the

Earth's magnetic field. At the exact South Magnetic Pole the needle stands on end at 90 degrees angle, as it does at the North Magnetic Pole, but in the reverse direction.

Much of the sledge and trail training was outdoors, weather permitting, for both men and dogs – the fleet of sledges was kept tethered by ice-axes behind the hut. One night in early March, as they sat at the evening meal, the wind rose to hurricane force, scraping and sloughing the frozen snow from the roof over their heads; suddenly they were assailed by the noise of bombardment, a hammering and clattering as though the timbers were being beaten by rocks.

They rushed anxiously outdoors; the sledges had been ripped by the wind from their anchorages and were flying across the roof into space. Men raced here and there recovering them from crevices of ice or against rocks; all except one. Said Mawson, ruefully: 'The last we saw of that missing sledge it was skimming across the frozen sea, heading through the blackness toward Australia.' In the next weeks they lost other things, among them a small boat they had roped to a tall ice outcrop. Ice peak and boat alike were flung into a seething turmoil of sea, which Mawson recorded as 'fearful to witness'.

Now the autumnal equinox was close; the length of day and night were equal, the sun was slipping north across the equator to bring summer to the northern world and the long winter night to the Antarctic. The days would run out quickly; the Ides of March were close and Mawson eyed the packed sledge standing ready and looked along the tongue of ice running south, upward toward the misty plateau and the mysterious inland. It was an enticement, and there were good reasons for a sortie before the winter closed over them. The tongue of fairly smooth ice offered the only passable highway to the plateau; on either side, east and west, the great sliding canopy of ice overwhelming the land broke into rolling confusion of ridges, fissures, and crevassing, which ended atop icy cliffs towering 200 feet above a frozen sea. The only possible safe highway had to be surveyed for the planning that would take place in winter for the spring journeys; the way to the great curving dome of the inland had to be marked with poles, strutted with guy wires, that would line the route for sledgers who could not see their footholds in the drift-covered ice.

For his companions on this first penetration, he chose two men to whom he had mentally already allocated leadership of two separate sledging missions in the spring – this sortie was part of their apprenticeship. One was Cecil Madigan: tough, a strong-willed Rhodes Scholar from Adelaide, Madigan had been studying in Oxford when Mawson invited him to join the company. The other man was very different in nature and character – quiet Bob Bage[4], the wise-man of the whole team, Melbourne graduate and a major in the Royal Australian Engineers, the expedition's astronomer and clock-master.

When the wind dropped suddenly, late one afternoon, they made their first dash, helped up the steep slippery

---

[4]Major Bage was killed by a Turkish machine-gun during the Gallipoli campaign in 1915.

ice slope by three supporters – the ever-present Xavier Mertz, Bickerton, the aero-engineer, and the expedition photographer, Frank Hurley, one of the bright spirits among the young Australians.

They made only a mile on that first attempt; at an altitude of 500 feet, they were overtaken by dark and snow drift, so they anchored the sledge, left it there, and walked back to the hut. At noon next day the March sky cleared as they dashed to the tethered sledge and slogged upwards to the brow of the ice canopy hanging over Cape Denison. They travelled five and a half miles inland, and reached a height of 1,500 feet. There they topped the rise to look for the first time on the great wan inland wilderness, a vast, featureless icescape – running southward to the bottom of the world, dissolving into chilled haze. The gloom of the place filled them with awe; they stopped and erected a flagpole, froze in its footing and a temperature recorder, and then camped their first night on the plateau. Madigan made it memorable by sleepwalking and waking Mawson in the cold night seeking a wire pricker to unblock the jet on the primus stove.

When they opened their tent, the wind had lifted the snow to waist-high drift across the ice slopes. There was no point in going on to plot a safe path when the surface was not visible. They waited until midday, and when the wind rose still stronger, they packed the sledge, left it anchored and walked the five and a half miles back to the hut. It was a wise decision; that night of 8 March a hurricane blew steadily at eighty miles an hour, for several hours, and the temperature dropped to twenty-seven degrees below freezing. And from that day the wind attack grew increasingly fierce.

Before the autumnal equinox on 20 March, Douglas Mawson recognised high velocity wind as a common feature of the Adelie Land climate; he conceded he had brought his main party to 'the windiest corner of the world.' What they were to suffer in the next months underwrote his view and provided a record of great winds beyond human experience. On the eve of the equinox, 19 March[5], the southern wind gave a foretaste of its power. A sudden quiet stillness hung over the hut in the late afternoon, and then, in early evening, a great, distant, seething roar came to their ears. It seemed both to be to the south and at the same time overhead. Away on the ice headlands they could see great gouts of fast-blown snow hurling horizontally out to sea at about 600 feet above the cliffs. Further out, the bay was a seething madhouse with wild waves leaping into spume. The great gale was passing over their heads like a rumbling train of wind. Suddenly it fell on them and struck their home with a volley of squalls that rocked the hut on its foundations; its force was so fierce it drove snow powder through the tongue-and-groove boarding. Over the next days they felt stronger buffeting; on 22 March the hurricane blew steadily for over an hour at above ninety miles an hour, bringing heavy snow.

---

[5]This was the evening on which Captain Scott and Wilson and Bowers reached their last camp on the Ross Ice Shelf.

It was heroism, bordering on recklessness, to service their outside instruments in such blizzards. Only by bending and wearing the long-toothed Swiss-type steel crampons on thick leather shoes could anyone keep his feet. More often they crawled on hands and knees for safety against sudden gusts. Sometimes in the smothering snow-drift men were lost, only yards from the tunnel opening, or even standing on the snow-covered roof.

In the days of autumn, rising winds rushing from hundreds of miles inland accentuated these problems, bringing, as they did, the convecting armadas of snow clouds, which blotted out sun and sky, and diffused light so visibility came down to only a few yards. The flood of wind and snow caused Mawson to set down the facts of the environment and led him to add:

> 'The actual experience is something else.
> Picture drift that blots out the world, that
> is hurled, actually screaming with energy,
> through space in a 100-mile-an-hour wind.
> When the temperature is below freezing.
> Those are the facts. But, then shroud these
> infuriated elements with polar night and a
> plunge into such a black-white writhing
> storm is to stamp on the senses an indelible,
> awful impression seldom equalled in the
> whole gamut of natural experience...'

He said the world became a void: 'fierce, grisly, appalling; a fearful gloom in which the merciless blast was an incubus of vengeance that stabbed, froze and buffeted intruders with the stinging drift that choked and blinded.'

Sledging or travel would have been madness and they concentrated on their building program and preparation for better conditions. They even tried to erect their radio masts – but the great winds prevailed and they were not to transmit or receive any morse messages until the next year.

In April, the blizzards raged with only occasional lulls; on some days the gales rocked the hut with steady blasts above ninety-mile strength. On into May the great winds roared.

The winds on the night of 24 May included bolts of air that battered their hut so the roof vibrated and trembled and the hut shook on its foundations. They feared its sudden ruin and freely discussed plans of taking shelter in one of the snow tunnels. Extra struts and ropes were fixed to the rafters, but the winds cost them their sleep and gave hours of concern and anxiety. In the morning, Mawson examined the instruments and recorded that there had been blasts in the hurricane above 200 miles an hour.

At the end of the month the instruments showed that the average – or mean – wind strength over the whole of the days of May was 60.7 miles an hour. With the wind normally blowing about ninety-five miles an hour, men attending the outside recorders, or cutting ice for melting to drinking water, suffered frostbite. The cold fell to as low as sixty degrees below freezing.

Mawson noted that over many days the cold intensity varied only slightly. From his scientific training he reasoned that this steady intensity of low temperature was from the constant stream of air falling down from the high polar plateau where the reading might be expected at or about eighty degrees below zero. As this falling cold air flowed toward the sea it became compressed, and, so, rose slightly in temperature on the descending gradient. It brought a river of falling air over their quarters.

On the same coast, somewhere, were his other men, Wild and his companions. What had they found? Where were they? His own party was lodged in the chilled breath of this vast polar wilderness where he wrote: 'The winds have a force so terrific as to eclipse anything previously known in the world. We have found the kingdom of blizzards. We have come to an accursed land.'

THREE

The Winter of Intent

Berserk blizzards hammered the hut with bolts of air and rocked it as though under attack by rampaging fiends from hell; at times in the long, dark winter the great polar gales sweeping off the ice shield would lift suddenly, hundreds of feet overhead, and go rushing, roaring, carrying vast rivers of blown snow, which fell in shutes into the fuming turmoil of the sea. Strangely, sometimes, the howling fell away to a dead calm, bringing a silence that was weird, that always stirred anxiety, suspicion. Still, every lull saw the men stream from the hut to attack their outside tasks – and this included futile attempts to rig sixty-feet high wooden radio masts to try to meet the deep longing for some link with the world they had left behind. Mertz and Ninnis would harness dogs and run them up and down, cracking the whips, teaching obedience to commands – and the dogs were never happier than racing in teams across the snow with the sledge bumping behind and the two men yelling commands or singing student songs they'd taught each other. In an hour or two, in half a day at most, the wind would resume its force and the process of preparation would switch to the hut.

The living quarters were always a scene of action: men kept busy to keep warm, to avoid the oppressive gloom that indolence induced. The hut itself was not warm. Wintry temperatures laid a lining of ice on the hut walls with the benefit of excluding cold draughts and fine blown snow that had penetrated chinks in the tongue and groove boarding. It also edged bunks with inches of ice from breathing and the moisture of cooking. If the inside air went much above freezing, beds, bunks, men were plagued with cold drips and the frozen beaten snow trodden into the wooden floors dissolved into muddy slush.

This situation aided Mawson. Busy hands, in his philosophy, meant healthy minds and fit bodies. In this frozen setting was an ideal environment for his intent through the winter – readying equipment and men for the spring journeys... even to making sorties on to the plateau in late winter. He tolerated no idleness, and nothing, neither cold nor wind, interrupted the scientific program nor delayed the preparations.

It demanded physical steel, iron nerve, to wrap up and step out from shelter into a ninety-mile wind, blowing cold at about twenty degrees below freezing. Emerging in the dark of winter's night men were often at once bowled over – Mawson among them – and rolled in the wind until they struck some prominence or their crampon spikes took hold; sometimes, men stood at angles of forty-five degrees, leaning into the steady gale, to cut the ice for melting for their scarce supply of drinking water – in a world of frozen water.

At times, tough New Zealander Eric Webb crawled on hands and knees to the magnetic house there to sit in the light of a storm lantern scribbling notes on the magnetic fields, on the link with the flaring aurora that betokened magnetic storms from the sun.

Mid-winter came with the knowledge that the sun had halted its slide north of the equator and would now move back to bring spring to the southern hemisphere; and it was traditional Christmas for Antarctic explorers, celebrated with their last frozen chickens, wine and nuts and crackers – and not an isolated event, for, in the confining tedium of hut life in winter, the men looked for events to celebrate, even down to marking the anniversary of the first gaslight in the city of London. In the same way Xavier Mertz marked the Swiss national day, producing a hoarded box of 'perks' he'd carried from Basel for this event: pate de foie gras, canned Swiss soups; truffles, and a Bearnaise sauce for a main dish of roast breast of penguin and bacon. He was awarded the title of master-chef and accepted his gastronomic popularity with broad smiles. The last day of July also brought him happiness – he was allowed to hitch a team of dogs, during a lull, to lay a depot of food and stores on the ice face, and then to indulge in ski instruction for the assembled company.

That same night, with a ninety-mile wind overhead, the nightwatch recorded 'a wonderfully bright moon hanging like a great silver globe in the sky'. And Mawson wrote of seeing star-gems more lustrous than anywhere else in the world.

Supplanting all other beauty, bringing more hope and excitement, was the promised return of the sun – growing, extending with each day passing. July did not die easily. It left a battering memory, an assault of wind of ninety-five-mile strength – continuously – for seven hours, including bursts above 150 miles an hour. August brought hope. Yet, it was another ten days before Ninnis hitched the dog team into a sledge and went with Mawson and Madigan up the narrow ice-road to the brow of the plateau. There was a fifty-mile head wind and the dogs gave trouble, hating the sting of drift blowing directly into their faces, eyes and ears. Ninnis kept them working and, in a long struggle, the three men topped the rise to locate the sledge they had left, penned down with ice-nails, some five months earlier. It was on the ridge, five and a half miles from the hut, from where they'd raced home in a blizzard. Hard snow had compacted a ramp on the windward side and the snowflakes in the screaming winds had scoured wood, and eaten into leather and polished exposed metal. A thermograph, left iced into the surface, had worked for eight days, recording a low of sixty-seven degrees of freezing.

With the sledge they found a half-eaten fruit pudding, dropped in a hurry in the autumn. They put up their tent and thawed the pudding out on the primus and found it delicious, its flavour quite unspoilt in nature's deep-freeze.

Mawson saw this spot on the ice brow as his backdoor, a jumping-off spot into the desolation of the great ice shield. It would also be a refuge for men fighting their way back from weeks on the open trail, if they could be sure to find the location. Yet, while a strongly strutted flagpole might survive here, a tent encampment would not last a week – it would be shredded to ruin. The place bore the brunt of the great plateau gales before they hurtled downhill to Cape Denison, out of sight behind the rolling canopy of ice. There was an obvious answer; if there was to be an outpost to winter quarters it would have to go. They would carve a sanctuary in the face of the ice-cap.

–

The hours of 11 and 12 August they spent hacking a vertical shaft into hard blue-green ice, and then carving a room back into the solid frozen mass. Some of the ice they excavated built a sheltering wall for the snow-buried huskies; the rest went through a crevice in the back of their head-high, eight feet by seven ice-hole. It seemed a marvellous place to them; the silence was glorious, for the boom of the wind was now overhead.

It was large enough to accommodate four sleeping men; the opening at the back was ideal for waste disposal – it dropped into a crevasse reaching down to rockbed; and they could hear the ice creak round them. With their ice-axes they hewed out shelves for their food bags, cooker, mugs and possessions; the ice crystals spangled their helmets and clothing, and the light filtering down through the blue-green walls and ceiling caught all surfaces with brilliant facets of colour. It was magical, if cold. Ninnis was entranced. He turned to Mawson, pleading: 'Please, D.I. – let's call this depot something other than the five and a half-mile post! Let's call it: Aladdin's Cave!'

Next day, Mawson attempted to advance still further south, into the plateau country. The dogs, with the drift blowing into their faces, tried always to turn back. They made another two and a half miles before discretion ruled. They faced confused ice, a knee-high drift, and black, ominous cloud belts moving north from the polar regions. With only limited food and fuel aboard it would have been reckless to be caught out in a long-lasting blizzard; they turned tail in the fading light to race for their ice-hole. The dogs pulled badly – they tipped the sledge over twice, so Ninnis set them loose and they scampered ahead into the growing darkness of the short August day; only one dog stayed with the men, a special favourite of Mawson's, named Pavlova.

They plugged on through the murk until, suddenly, the flagpole at Aladdin's Cave loomed ahead. The dogs were there waiting in the snow. The sledge was unpacked and sleeping-bags, food, and the cooker-box were passed down the shaft. Ninnis fed the hungry dogs with seal-meat and biscuits and then covered the entrance to the cave with the tent. The place was again magical to him; Mawson was bent over the primus, and the hissing blue flame lit the ice-hole with a warm look of cosiness and added to the fragrance of the traditional sledgers' 'hoosh' – a mixture, boiled in melted snow, of pemmican (dried beef, heavily laced with powdered fat) enriched with a knob of butter, and thickened with dry biscuits, which were beaten to crumbs with a geologist's hammer. Out of the wind, in the stillness, well fed, they slept undisturbed.

The blizzard held them there for two days, then Mawson, aware that their absence would arouse anxiety at the hut, decided to make a run down the slope in the thick drift, using the east-west cracks in the snow and the polemarkers to find his way. So steep was the slope in places, the dogs were a hindrance, tipping the sledge over with too much speed. They were set loose, Ninnis assuming they would run with the sledge or go back to the hut. The three men fought on down the ice road as the drift rose above their heads, stalking their way and holding back the sliding sledge by digging their crampon spikes into the ice.

They reached the hut in the dark of evening to find the dogs had not arrived − all six were missing. The gale rose into a ninety-mile blow. It lasted all next day and prevented a climb to the ice brow. Mertz and Ninnis made two attempts on the following days but failed after travelling no more than a thousand yards. For four days the blizzard held them to the hut. Then Mertz, helped by Bage and Hurley battled upward through the whirling drift in a sixty-mile wind, to locate the flagpole at Aladdin's Cave. All the dogs were there. They saw Castor leaping up and down, excited at their coming; Pavlova was in a good state, but the other four were apathetic and sick, especially the one called Grandmother for an appearance that belied his sex. The men took the dogs into the ice-hole, cooked hot food for them and nursed them for two days. Alas, Grandmother did not recover and they buried him outside in the ice. After three days they made their way back to the hut and the dogs were given further recuperation.

On the first day of September Ninnis and Mertz took a team to the ridge to deposit bags of food for the coming journeys. They allowed the huskies to run loose and then tried to coax them to run back to the hut with the returning sledge. Aladdin's Cave, however, was some inexplicable magnet for the dogs. Mertz harnessed them and led them down the slope, but, when loosed, they again all ran back to the hole in the ice. Again, they were put on a trace and taken in line all the way back to the hut. That night, when Ninnis checked the charges, two dogs had found a way out of the tunnel by digging through the snow and had run back up the slope to the plateau ridge. These two errant animals were named Scott – and Franklin. Told of their behaviour, Mawson shrugged his shoulders: 'The ways of Greenland husky dogs are beyond human comprehension,' he said. The two missing animals were located at the ice-hole; Franklin, welcoming the men with wagging tail, was snuggled down behind the sheltering ice-wall; the dog, Scott, the wildest of the two, slunk away like a wolf. He was never seen again and it was presumed he fell down a crevasse and died.

–

In the hut at Cape Denison the tempo of life underwent dramatic change. Preparations became more evident and an air of suppressed excitement spread through the company. Men spent hours at the main table weighing, measuring, packing the food bags for spring journeys. All expeditions were to be based on teams of three men, and food allotments were made in allowances of pemmican, sugar, biscuits and the like for three men for a week. Then, each weekly bag was stored into strong, larger canvas bags called 'food tanks', with a month's rations for three men in each tank. Along with this work, Mertz and Ninnis spent long hours over the stove, drying out seal meat for the dogs. Weight on a sledge had to be reduced by every possible ounce, so, men carried pemmican, cocoa, tea, reinforced chocolate. For the dogs there would be dried meat and biscuits.

In mid-September, Mawson pushed ahead with plans to lay food depots along the trails leading beyond Aladdin's Cave. One party (with Eric Webb leading to gain new magnetic readings) pushed south for twelve miles, where they cut a new ice-hole called the 'Grotto,' and then were virtually blown back to Aladdin's Cave. Mertz and Ninnis took a dog team to lay a food depot in the south-east, while Madigan and two companions pushed into the virgin west – with him were two as yet untried sledgers, New Zealander Dr Whetter, and Australian biologist, John Close. Madigan was a determined man. When two weeks passed without a sign of this party, Mawson's concern grew into anxiety. After sixteen days of absence, Mertz and Ninnis drove a dog team up to Aladdin's Cave, and there, came on three gale-beaten, worn, and tattered men. Fronting eighty-mile winds, they had fought their way back from a penetration of fifty miles into the west, meeting steep ice slopes rising to 4,500 feet. The wind won in the end. Hours of sewing with frostbitten fingers, patching rips in the tent, came to nought on their way back when they were thirteen miles from Aladdin's Cave. There the hurricane ripped the cap from the tent, snapped the poles, and collapsed the whole structure over their heads. They made a forced march through the snow and fog, and, by great fortune, found the flagpole marking the ice-hole, the white scars of frostbite on their faces.

Telling his tale in the hut, Madigan looked at Mawson and said: 'D.I.! All that toil cutting the cave has paid off, I reckon. Without question, it saved our lives.'

The western probe helped their preparation. It was blindingly clear that the normal type of summer tent, used on the Scott and Shackleton expeditions, would not suffice against the winds of this terrain. Tents, and clothing, would have to be strengthened and adapted. The usual method of erecting a tent — by setting the frame of five bamboo poles into the snow and then billowing the tent cover over the frame — was abandoned. Mawson had the poles sown into the tent cover which was lined and strengthened — with hours of stitching in new canvas, on the treadle machine. The skirts of the tents were extended; it was planned that the tents would be erected like umbrellas, with one man inside and two men outside holding the structure against the wind. Once the tent was open the poles would be dug into snow, or into pre-cut holes in the ice, and blocks of snow or ice, cut previously with the pickaxe or spade, would be piled on the skirting to hold the cover firm, to prevent the wind getting under the tent and lifting it from over their heads. Tents were vital to their lives on the trail, he told his men. They had to be proofed against savage wind, but they could not weigh *too* much, since they had to be dragged every foot of the way on the sledge.

He was meticulous over clothing. In the savage Adelie Land weather the most vital garments were the outer wind-proof jackets and pants of a material they knew as burberry – tough, lightweight, close-woven material made into roomy garments that kept out snow, and water, as well as wind. The trousers were double thickness to meet constant contact with snow – they came well above the waist and were tied with braces of lampwick, the Antarctic sledgers' standby: soft, easy to untie and, unlike cord, lampwick added to warmth. The jacket, usually with a hood attached to be pulled over a woollen helmet, was tied at the waist, again with lampwick. Hands were protected with fingerless woollen mitts, under full mittens with one section for all four fingers for greater warmth, and wolfskin over-gloves, again tied round the neck with lampwick to obviate the risk of them being blown away and lost.

Their footwear was of two kinds; around the hut or on short winter journeys they wore heavy leather boots to which steel crampons could be attached, such as those worn by mountaineers; for the weeks on the open trail they took to the warmer, lighter finnesko: fur-boots of reindeer skin lined with layers of the special, dried grass from Lapland called sannegras, or with Manilla fibre, which, with several pairs of socks, gave greater warmth and ease of walking on snowy surfaces. A lightweight type of crampon could also be worn with these finnesko.

Sledges, from both Norway and Australia, were eleven or twelve feet long: the Norwegian sledges were lightly built of ash and hickory; the heavier Australian ones of mountain ash or spotted gum.

Thus, the experiences of Madigan and his companions on their 100-mile journey induced new activity into the daily routine: the furnace glowed, the grinding wheel shot sparks, men fashioned innovations in crampons while others cut and stitched – tents, burberry coats, and pants, extra bands of canvas on their finnesko to stop snow creeping up into the legs of their pants; there were alterations to ropes and harnesses. All this while the preparation of food and training of dogs took place. Mertz and Ninnis, as well as feeding and cleaning the dogs daily, also filled the hut with the smell of drying seal meat cut in frozen blocks from the walls of a snow-tunnel; and they tackled another task.

Using needle and thread, and, when it was available, the treadle sewing machine, they made individual harness to fit each dog. Experience on their few journeys had shown that the general harness was ill-fitting for some animals, that, by wriggling vigorously, they could slip free – a dangerous condition for both dogs and men on the open plateau. From rolls of canvas strip and leather belting, Ninnis and Mertz cut and fitted the eighteen animals. It was a cold, finger-testing task for Ninnis and Mertz. Yet, it was done carefully, even tenderly; for, to these two men, the huskies were their family. Some of them they had nurtured from pups, who had been born of parents who died on the voyage. Most of the dogs they had named:

John Bull, Basilisk, Caruso, Scott, and Shackleton, and Gadget who produced the fondled, spoilt, beloved pet pup of the hut, little Blizzard. Other pups had been born, but in such conditions the mortality was high, even though grown huskies could snuggle down into snow in the worst winds and sleep until hunger moved them, or – as often happened on the trail – the drift in their ears and eyes and noses made them howl with discomfort.

In the isolation of his own compartment in the hut, Mawson faced the final decisions on the pattern of his Antarctic exploration. September was flitting away; men and dogs were on the edge of readiness; at any time good fortune might bring fine weather and give them months of travel before the *Aurora* returned around mid-January to take them all home... but just how long would this accursed climate give them?

Full surveys of the 2,000 miles of coast and its hinterland that was immediately beneath Australia was now a pipe dream for two bases separated by – how many hundreds of miles? The best that could be done was to stay with his fanwise method of operation and meet his undertaking to make a better scientific job of locating the South Magnetic Pole. That was the first party to be appointed.

Eric Webb was the magnetician; Bob Bage, steady, a good all-round technologist, would lead the trio; Frank Hurley, tough and energetic, would go with them to take the first photograph of their finding the location of the actual magnetic pole. Theirs would be a hard, south-south-east journey, up and over the rising plateau, the most southerly of all their exploration.

With the understanding that Frank Wild – somewhere in the west – would drive eastward, Mawson decided to send a group of three in that direction to cover as much coast and uplands as was humanly possible. This party would have the aid of the aircraft, as an air-tractor, pulling a loaded sledge over and beyond the land on which Madigan had reported. Bickerton would head that three-some, with Dr Whetter and young Arthur Hodgeman, the government map-maker assigned to the expedition for that purpose. Both these parties would have support at the start of their travel from men who would come back to give support to other parties, or to man the hut in the meantime.

Mawson planned a concentration in the east. Explor-ation in that direction was all the more important, in his view, because if the cruise of the *Aurora* was taken as a guide it was a coastline not likely to be surveyed and visited by ship. Indeed, he decided, three separate parties would go in that general direction, along the coast and its hinterland, which he already knew contained the greatest glacier, then known, on Earth.

One party, formed partly from other support groups, would scout the nearby shore region east from Cape Denison. To lead this trio he chose Dr Frank Stillwell, a fellow geologist. Another shore-line trio to leapfrog the first party would be led by Cecil Madigan while he himself would lead the toughest of all the journeys, the deep probe across country into the unknown territory running back east, toward that point he had reached in 1909 with David and MacKay when they crossed the Prince Albert Mountains into what had been named Victoria Land. In his mind he hoped for a journey thrusting 500 miles into the *terra incognita* he'd dreamed of walking when on Erebus. The far inland called him, but he wanted to stay near enough to the coast to plot its geography for map-making, for claiming the terrain for the crown, and yet to be able to move far enough inland to find the mountains, perhaps to find the moraines of glaciers that would reveal clues to the geological nature of that part of the continent or the mineral treasures that lay under the overwhelming ice shield. There could be great wealth at Australia's back-door.

It was, of all the journeys, the one that would need the dog teams – all that were fit and available – and that fact at once selected his two companions... the quiet, strong 'Cherub' Ninnis and his sturdy friend, the ener-getic ski-master, Dr Xavier Mertz. In all the trials, out from England with the dogs, the landing, the winter, they had shown their mettle. He could not have better companions.

If all went well, the parties would explore about 1,500 miles of unknown coast and its immediate hinterland – a greater area of exploration in Antarctica than attained by any previous expedition.

He announced his plans to the men and they held free and open discussion. Food, fuel and shelter, he emphas-ised, were key factors; their diet was carefully balanced for the heavy physical demands of man-hauling sledges; they had ration packets that gave them just above two pounds of food weight a day, but that was dried food to which water had to be added. To those who had not yet savoured the savage hunger after a long day hauling a sledge, he said, the hot 'hoosh' of dried pemmican and butter in their mugs would be a most delicious meal. With it would go chocolate, biscuits – of the hard, dry variety – sugar, cocoa, tea bags and some dried milk; there would also be the additional 'perks' of raisins or sultanas, which each could carry in their personal pack.

–

The days of September brought little relief from the wind. October, and again the days offered little hope to a start of serious sledging. When the lulls came, they foraged among the ice hunks, seeking marine biological specimens, their eyes always turning to the sky, looking southward over the great rolling brow of distant ice. The waiting became a trial; any event that broke the tension was joyfully embraced.

Blustering gales swept the high uplands and brought deluges of snow overhead, shutting off their view of the bay with curtains of falling white drift; the cold and the wind seemed little different from winter. The only difference was daylight.

Then, during a lull in the late afternoon of 11 October, Dr Archie McLean walked to the edge of the ice to be startled by a short dwarf figure waddling ashore, on to the ice ramp, out of the heaving water. The first penguin of the season had returned, and had come ashore, as equally astonished as the human being greeting him. He was to be even more bewildered. McLean clasped him in his arms, and calling excitedly to his comrades, hurried the creature into the hut where he set him down on the mess table. 'Harbinger of spring,' he triumphed, 'This simple soul is all our daffodils, crocuses, primroses, tulips – all rolled into one!' The men caught the spirit; wine was produced to celebrate the occasion of the coming of spring and the penguin. Only the cook grumbled – because the startled bird left its mark on the mess table.

From the day of the first penguin, Mawson watched the chill world come to life again. Thousands of penguins struggled out of the heaving bay to find their old haunts, to build their pitiful little nests of rings of stones; the seals lunged and lolloped onto the rocks, beautifully fat, sleek creatures, cousins to the prized and decimated fur seals already hunted near to extinction in southern waters. As he watched them come back, with the first flights of birds winging in, the surge of returning life lifted his heart and gave a glow to the last weeks of waiting for the first marches of exploration. He found in himself a deep affinity and a lasting affection for all forms of life that struggled hardily against a savage environment; all his life he would oppose the slaughter of seals for furs, hate the hunt of whales for blubber, and condemn the massacre of penguins for a miserable yield of oil.

Yet, there was still no glimmer of hope of travel on the open plateau. That month the average wind blow – over all the days – recorded a mean of fifty-seven miles an hour. Then came further indications of the southern quest of the sun; at the end of the month the ice lining in the hut began to melt. The living quarters were filled with tin-cans hung on string to catch the drips, the walls lost their frozen glaze; the floor became mushy.

With the higher temperatures Bickerton called a corps of volunteers and the Vickers aircraft fuselage was laboriously dug from a mountain of snow, and there were days of patient thawing before the first aero-engine coughed to life on the Antarctic mainland. A mechanical bird with broken wings, it was fitted with skis and make-do brakes, to pull Bickerton and his men to the west.

So, they came into November 1912, with the continent still unyielding and Mawson's concern growing. The old ship would be back by the second week in the new year. If they got all the parties away in the next week – how far could they hope to travel? His own target was a spot some 500 miles to the east. Bage might have to move a similar distance to locate the Magnetic South Pole. Could they expect as many as sixty days of travel – and to trudge an average or more than fifteen miles a day? The arithmetic was so tight, every day of marching would be precious. He could not guess at the conditions to be met, once they were on the plateau, moving west and east. It was another crisis of timing that faced him; to be back in time for the ship.

He told them of his thoughts at the mess table on the evening of 3 November. He said: 'Men, we just can't wait any longer! Everything has been done for successful journeys, but we court defeat if we wait too long for favourable days. I say we must go soon. We must be on the open trail within a week. The *Aurora* will be back on the first week in January, all being well, and then we go to pick up Frank Wild and his men. Whatever the weather, we go in the next week… and remember, we must all be back at the hut by 15 January. At all costs, we have to be ready to go home by then!'

FOUR

Outward Bound

In the cold, black night the hut was a cocoon of light, warmth, colour, and laughter. On this evening of 5 November, the smell of roast mutton and hot fruit pudding gave way to the fragrance of coffee, drunk with port, of oranges and cigar smoke. The great mid-winter feast, the memory of Mertz' culinary surprises for the Swiss national day, the birthday dinners and celebrations all paled in retrospect against this repast – and the event that it commemorated. Under the bunting, the draped flags, they linked hands for Auld Lang Syne and drank to the King.

This was the farewell. The day they'd trained for, suffered for, longed for, was at hand. Tomorrow they would start the march into untrodden country.

There was a boisterous impromptu toast to the expedition leader: 'To D.I. God bless him!' And Mawson, responding, toasted them all and his two chosen companions and said: 'God speed you all and bring you back safely to board the *Aurora*'... and then they toasted the ship, and 'Gloomy' Davis...

Ninnis remembered the words when they all turned in. He was heady with wine, replete with food, luxuriating in the woollen blankets and the hermitage of his wooden bunk. What would there be out there? Weeks of sleeping in a skin bag on the frozen snow and ice, sharing the tent and the food... where would they go in that cold unknown? He heard the wind from the plateau scouring at the bitten snow on the roof of the hut.

The vilest weather intervened. Morning brought a swirling blizzard that blotted out the world to a few yards of visibility. Movement onto the narrow ice-slope would be to invite disaster. With the sledges standing ready, they played cards or chess in the hut, and the whole day went away with no hope of a dash to Aladdin's Cave. The gale continued blustering, and filling the air with snow, all through the night and into the next morning.

Not until noon on 7 November was there a fall in the tempo of the wind. Mawson sent the first support party away; a sledge, loaded with food tanks and fuel, spare tent and tools, was towed by Dr John Hunter, Bert Murphy, the storekeeper, and Charles Laseron, biologist. They were cheered off in low drift, the little flag on the sledge mast fluttering wildly as they slogged the steep slope behind the hut. Then, they were out of sight.

A half-dozen men filled the rest of the afternoon collecting eggs for the larder from the nests in the teeming penguin rookeries... then the wind came battering down from the slopes, and with their boxes of eggs they struggled to the hut, almost bent double against the renewed onslaught. Two hours later the first support trio were back in the hut. Their loaded sledge was tethered to an ice-pick, two thirds of a mile up the ice-road. There, a head-on wind grew so strong they could not make further headway.

There were glum faces in the hut that night; men cursed the terrible climate. Mawson tried to cheer them. 'It is still early summer. It could get a lot better and the more rest you get now the further you will march when the conditions are good.'

Next morning proved his point. Before noon he had sent three parties on their way. The support trio marched to the pegged sledge and towed it to Aladdin's Cave, and after them went the two east coastal parties who would meet with Mawson, some twenty miles out from the ice-hole, with a few days' extra food for him to take aboard for the long far-eastern trek. The Magnetic Pole team was to leave with his own trio in the morning; and they would travel together to the cave.

The hut seemed quiet and empty with half its occupants gone. Bickerton, the only party leader with no immediate prospect of a start, was almost biting his nails in frustration, wishing all the time for a sudden rise in the temperature so he could be sure that the engine of the aero-tractor would continue to fire[6].

---

[6]In the event, Bickerton's western party did not get away until 3 December. Then, the engine pulled their sledges for forty miles and stopped for ever, to be a hunk of metal and wire in the snow. The first aircraft in the Antarctic was crippled with a broken propeller and frozen engine block.

Again, the wind rebutted Mawson's planning. November 9 became a doleful Saturday, men sitting disconsolate in the hut with the air outside filled with thick drift. Mawson reasoned the same weather would hold back the exodus of the other parties at Aladdin's Cave, and that a dash up the ice-slope would cause congestion and be dangerous. He took time to write a letter to his betrothed, although there was no hope of it being delivered until the *Aurora* came back and took them home. He was moved to write of his 'disappointment in this land', and to say: 'I am writing this note in case anything should happen which will prevent me from reaching you as soon as the mail from here, which is expected to be picked up next January. So many things may intervene, for, one truly lives from day to day here, and our sledge journey is about to commence.'

They were up early on Sunday; the white chaos still blew down from the plateau. Mertz dispelled the gloom at an insistent suggestion from Ninnis. A box of penguin eggs was brought to the cook's table, a small block of blubber was thrown on the dull fire to spark it to life, and an apron was tied on Mertz with orders to produce breakfast. Eyes shining, strong teeth flashing in the broadest smile, he announced: 'We shall a wonderful breakfast make – and then the sky will clear.' Deftly he beat the eggs into a golden foam, spiced with his own selection of herbs and wine, and cooked this into delicious omelettes in pans steaming with melted butter. Mawson delighted in watching Mertz in his central role as the popular master-chef. It was a glowing memory to take onto the ice.

At ten that Sunday morning, Mawson held a brief service. He read from his prayer book, they sang a hymn and sought divine help for their safety and for clement weather.

By noon the sky cleared. Mawson added a hurried postscript to his letter to Paquita Delprat:

> 'We shall get away in an hour's time. I have
> two good companions, Dr Mertz and Lieu-
> tenant Ninnis. It is unlikely that any harm
> will come to us; but, should I not return
> to you in Australia, please know that I truly
> loved you. The others are waiting.'

Outside the hut Ninnis and Mertz stood ready with the seventeen dogs and three sledges. The huskies were yapping and straining to be gone. Somewhere deep in their atavism the Eskimos had implanted the love of running in a team, pulling for man. Their eagerness to be in harness, to be off, was a pleasure to witness. Ninnis had his own cry to replace the usual 'Mush! Mush!' A crack of the whip and he shouted: 'Hike! Hike!' – and they were off, onto the ice shield, racing up the slope, the teams straining to catch the Southern Magnetic Pole party of Bage, Webb and Hurley, who were man-hauling their own sledge to the ice brow. The five and a half miles were made in four hours in bright sunshine and almost calm weather. They shared a meal at Aladdin's Cave with the Magnetic Pole party, who, anxious to catch their southern support team, then hurried off to the south east, over the rolling inland ice field.

While Mertz and Ninnis tended the dogs, Mawson strode to an ice hillock and scanned the east with binoculars. In the low, south-west light he could see how the folding downward spread of the colossal ice canopy shattered its own surface into ruts, ridges, and broken fields. Black-etched, deep lines along the eastern horizon spoke of dangerous crevasse conditions, to his mind. There was no direct path for him to the east. He would first strike south for several miles to climb the hump of the ice-cap, and then turn east. He had a rendezvous to keep with the two coastal parties – who must have travelled the same way as he was now forced to follow. They were to meet some eighteen miles out from Aladdin's Cave where he would take aboard the last extra rations and fuel from their supply; each extra day's supply of food for men and dogs they could take on would mean deeper penetration into that far land.

He had started with a heavy enough load, and more would be taken on here at the Cave depot; but, two thirds of all their load was food so that each day of marching cut into it. Their load on leaving Aladdin's Cave he had calculated to the last ounce… a total of 1,723 pounds, including the three sledges. Of this, food and fuel accounted for 1,260 pounds… and 700 pounds of that food weight was dried seal meat, blubber and biscuits for the dogs. Every mouthful was essential; food was energy, and energy was distance. Six one-gallon cans of kerosene, weighing sixty pounds – with a few bottles of alcohol to prime and start the primus stove – would last their needs for cooking… and for drinking melted snow.

The rest of the weight they had to haul – some 463 pounds – comprised items essential to life support and to the needs of their scientific mission of exploration. Their strong canvas tent weighed forty-four pounds with its five bamboo poles sewn into the lining; there were three reindeer skin sleeping-bags – a total of thirty pounds, and there were the ice-country tools, pickaxe, spades, the geologist's hammers, the surveying equipment, theodolite in its box, and its tripod, prismatic compasses, altitude meters, thermometers, binoculars, camera and films, ropes, harness for men and dogs, and cooking equipment, which was in the main box – the aluminium Nansen cooker of inner compartment, in which their food was heated, and an outer section where the snow was melted to make hot drinks... together with the primus stove and its repair kit. There were bundles of spare clothes, extra finnesko, to replace those fur boots that might wear rapidly on rough ice, plus a bundle of the dried grass used to line the soles. There were rolls of lampwick for lashings, an extra coil of alpine rope; there were ski shoes and stocks and medical aids, books of tables and logs for positional determination; and there was a ground sheet, a .22 rifle and bullets, fishing lines and hooks, repair kits of several kinds, including a sailmaker's needle-and-thread outfit for patching clothes and tents; there was also a spare, lightweight tent cover, in green drill, to save wear and tear on the heavier tent when the weather was kind.

Each of the men was allotted thirty-two ounces of concentrated foods per day. The bulk of this would be pemmican and dry, hard, wholemeal biscuit; there was also cocoa, mixed with rich dried milk, and sugar, with minor amounts of cheese, chocolate, raisins, and tea. This was the food that had been taken from the original containers during the winter and packed in calico bags. All these were now in a large, waterproof drill bag, with a square base to fit the sledge platform and a tie-neck fastened with lampwick to keep out moisture. When Mawson took on his last supplies at the rendezvous with the two coastal parties, in the next day or two, he estimated the food would be sufficient for a nine weeks' journey.

Climbing down toward the cave he felt the wind flowing steadily off the ridge at about twenty-five-mile strength; a pleasant breeze for this country, he told himself. The dogs had been fed and tethered; they were buried in soft snow behind the sheltering wall. He read the temperature. It was then thirty-six degrees below freezing.

The cave was a cosy retreat – they had a storm lantern burning a soft yellow light reflecting gold from the icy walls. Ninnis and Mertz were in their sleeping bags, both reading. Ninnis had a much-loved miniature edition of Thackeray's novels, given to him by his mother when they had said farewell at their home in the London suburb of Streatham. Mertz pored over a Sherlock Holmes book – part of the English curriculum, which he had struggled through so often during the winter he could recite whole phrases. Mawson crawled into his bag and reminded them he wanted an early start in the morning.

Mertz looked at him, a half-smile rounding his cheeks: 'All the times here we have been, never was it so quiet and still. Perhaps, an omen, yes? I believe in omens, Doctor. Yet, I will a good prayer say for our journey and for fine weather.' Mawson scribbled his last notes in his journal. He quenched the lantern and added his own prayer.

It was seven when he woke; the morning was bright but it was still very cold. There were far layers of dark cloud southward toward the polar plateau.

They were fed, packed, and away south by 8.30. Mawson broke the trail ahead. The first dog team pulled two sledges carrying between them one half of the entire load; this was to lighten wear on the runners of these two sledges, which would come into fuller operation later in the journey. The second dog team, driven by Ninnis, hauled one half their entire cargo on the third sledge, the one that would be discarded when the food load was reduced.

The animals were excited, so eager to race across the rumpled surface they set the sledges rocking and jolting, and bringing them near to capsizing. Mertz loved the pace, singing student songs in German. But, Mawson called a halt. The bumping across the ice ridges loosened the sledge meter and strained the universal joint by which it was attached to the rear of the second sledge; he could not risk further damage so early in the journey. From there Mertz went ahead on skis as front-runner and Mawson drove the leading dog team. By noon, they had run about seven miles. Soon they turned east and the looming clouds blanketed the sky and brought heavy snow showers; the wind started to rise above forty-five miles an hour and the blown snow drifted higher than their knees.

They could not go on and so camped about eight miles out from Aladdin's Cave.

It was a dismal start to their journey. The ice surface was corrugated under their tent into small sastrugi, and the blizzard mounted over them with rising wind. They had trouble cutting blocks of compacted snow to weigh down the skirt of the tent; and the dogs howled in discomfort while the snow coated them until only their black noses were showing. The men could do no more than wait in their sleeping-bags. They accepted the credo of 'no work, no hot hoosh', on the principle that the concentrated food was to meet bodily requirements after a hard day slogging across the ice, that the same food, eaten while they were resting idly in their bags, would be nauseating. So, they nibbled dry biscuits and chocolate, occasionally boiled water for cocoa and tea, and, as Mawson said, 'saved the hot rations for a rainy day'.

All through the next day – Tuesday 12 November – they huddled down trying to keep warm as the wind slapped the tent cover against the poles. Their backs ached from the unyielding ice and sleep came only in snatches to relieve their growing frustration.

By noon on 13 November the air cleared sufficiently for them to strike this uncomfortable camp and to make a forced march to the rendezvous spot, eighteen miles out from Aladdin's Cave. They made the location by 7.30 pm, but drift was again rising and they could see only a few yards. Again, they pitched the tent, struggling to erect it against a strengthening wind; during the night the blast rose about eighty-mile strength and, at times, they all three clung hold of the vibrating bamboo poles to hold them down into the ice. As their shelter rocked and shook, Mawson feared the excessive sudden blast that could rip the tent from above their heads.

This blizzard buffeted them for three days. They had no hope of lighting a primus until the Saturday. It was a rigorous introduction to long-distance travel for Ninnis and Mertz. They both stood up to it well, although, when they crawled out through the cloth tunnel that formed the sealed entrance to their tent, to face the clearing scene on the Saturday afternoon, Ninnis felt faint and squatted in the snow for a time to recover. Mawson put his faintness down to lack of exercise. Mertz was not affected – he was soon digging out the snowed-in sledges and locating the dogs under the drift and cutting them free where their hair had become frozen to the sastrugi. He laughed to Mawson: 'They are such fun – they look like live snow-balls. But, they are so hungry they could eat me, I think, or any of their neighbours.'

Under the clearing sky the wilderness about them opened to a vista of blank white; there was no sign of the six men of the two coastal parties. Apprehensive that some catastrophe had overtaken them, Mawson unloaded a sledge, and, with a dog team and Mertz on skis, raced across the wastes to return to Aladdin's Cave; he had travelled a mile when Mertz's sharp eyes saw a moving black dot in the north west. Through the glasses this dot focused to six men with two sledges, and they at once raced back to re-join Ninnis, and, then, travelled to intercept the two coastal parties.

They made a convivial camp that night – three tents and nine men in good weather, eating their meal in the evening sun. Yet, it was worrying. They had been out almost a week and were still only eighteen miles from Aladdin's Cave.

The three parties travelled in convoy next day, parting in the afternoon and saying their final farewell. The two groups then moved north-east toward the coast and Mawson travelled due east on a rising ice field. At five in the evening, he topped this rise after a climb to 2,600 feet altitude. Breathless with effort, they stopped to look ahead at a scene new to man. Far to the north the ocean dissolved into a hazy horizon; ahead, open water barred his eastward march where the coast swept back south in a great indentation. Beyond were rolling ice hills and centred in the distance was the site where the great floating glacier they had discovered the previous January conjoined the soaring ice barrier of the southern land.

They turned their faces south and marched into rising, rolling ice hills, pocketed with packed snow, and the light lanced and bounced and banished shadows and shapes, and, under the overcast sky, blinded their eyes to the ridges and troughs. The sensation of contact with the continent was only through their stumbling, tripping feet in a weird, diffuse glare that penetrated their dark goggles and forced their heads lower. Mawson broke trail, determined to make a good march that day. When they ascended to level terrain, when he'd seen the sledge meter, he called the halt.

Now, they followed the ritual. The tent was spread, point toward the wind direction. Five holes were hacked into the ice for the bamboo poles to sit; then Mawson took the spade and cut snow blocks to weigh down the flounce — or skirting. Ninnis had sewn extra width to this to meet the savage drag expected from winds in these marches. When the snow blocks were cut, Ninnis crawled into the flattened tent cover, and, as Mawson and Mertz lifted the frame, he kicked the pole-ends into the cut holes, opening the whole like an umbrella, but at the same time holding it down from the pull of wind while the two men outside loaded the snow blocks onto the skirt. Ninnis then came out by the funnel-like exit — which could be tied shut, inside or outside, with attached lengths of lampwick. Mawson entered the tent, and Mertz and Ninnis unpacked the camping gear. They handed in the floorcloth, or waterproof groundsheet, and the precious cooker box and a bag of food. Then they fed the dogs, and tethered them with traces tied to ice-nails driven into the frozen surface. The sledges were anchored to an ice-axe, the sleeping bags unpacked and also passed into the tent. Mawson, by then, had the primus roaring — a testing achievement in such cold where the crystals of breath freeze on the ends of matches, and the alcohol primer is reluctant to burst into flame to heat the gas outlet... now the two compartments of the Nansen cooker were packed with snow, and it was slowly lowered onto the burning jet. If he was too hurried in his action, he could easily put the flame out — and the taint of kerosene would permeate their food. The snow in the centre compartment, being immediately over the flame, melted first, and into this now heating water Mawson mixed the heaps of pemmican, a generous knob of butter and a helping of dry, wholemeal biscuit, which he had crushed with his rock hammer. Three enamelled mugs sat on the floor cloth waiting to be filled. They did not use the system, sometimes followed, of the two men turning their backs and the cook, with one finger on a serving, asking 'Whose?' It was far too early in the journey for the savage bite of hunger to rouse suspicion of favoured serving of food. These mugs were equal. They could sit and watch as the hot, thick brew was poured.

'We have earned this,' Mawson said. 'We had a good march of fourteen miles; close to the average I want for the whole journey.'

Their eyes had suffered the assault of glare all day – white reflecting on white, light, hazy, sharp, blurred. Now it filtered, muted green, through the tent; the primus burning under the snow that had melted in the outer container was a brilliant blue. Ninnis looked at their different 'ditty' bags, Mawson's yellow, his red, Mertz's orange.

He said: 'Colour is suddenly precious, warming to the mind. It was thoughtful to give us these personal bags in such colours.' He touched Mawson's memory.

'Yes, Cherub. It was thoughtful.'

Two thousand miles north, in a different, soft, warm, world he'd seen Paquita and her mother stitching these bags. 'They'll give you a lift in that dull white world,' Paquita had said. They had sat together, mother and daughter, on a cane lounge – on that same verandah where on a gentle, summer night last year, with the warm, rhythmic sound of surf pounding the beach at Brighton, near Adelaide, he had asked her to marry him. 'After I come back,' he had said. Red, yellow, orange bags – Ninnis thereafter called them 'Paquita bags.'

Mawson, still thoughtful, boiled the melted snow in the outer cannister and stirred in heaped spoons of cocoa, mixed with sugar and rich dried milk. They smiled at each other as they supped at the hot, sweet drink, feeling its warmth run through their bodies.

They still had ritual chores: wet finnesko, socks, mittens, and woollen helmets were hung on a lampwick line tied to the tent-poles. These were carefully moulded into shape for donning next morning... they would freeze stiff overnight. Mertz and Ninnis unrolled their sleeping-bags and climbed in while Mawson packed the cooker box. Being cook, he slept in the middle and was last man into his bag. Both Mertz and Ninnis pulled the reindeer covers over their heads and fastened the toggles under their chins. Mawson sat up, first of all, and wrote notes in his journal. Then he closed the bag over his head, and lay back, with the hard ice under his shoulders.

The tent was just wide enough for three men in separate sleeping-bags ... shoulder to shoulder. It made for comradeship – and that was a heartening thought for this first stage of their outward odyssey. 'I have two good companions,' he had written to Paquita.

FIVE

Discovery and Death

From the first days of their eastern journey the forbidding land met them with implacable hostility. It imposed exertion, which ate at their reserves of strength and steadily sapped their stamina; it confronted them with heart-pounding, steep, slippery slopes, with suddenly precipitous declines and with impassable upheavals of ice. Its malignant winds scoured their faces and chilled their blood; and its light was a blinding, refracted glare, bouncing viciously back and forth, burning their eyeballs, and shrouding the pitfalls and perils under the coating snows, masking distance and size with cloudy distortion.

Succinctly, clearly, the frozen sixth continent said: this is no place for man! Yet, because it erected such barriers, the continent was spiced with the temptations of discovery and challenge...

Deceptively, the long march opened under a clear sky and a beaming sun – a sun so warm they spread their sleeping-bags, turned inside out, to dry away the moisture from their sleeping bodies; and then they trudged upward into the snow-covered ice with the eager dogs yanking at their traces, and their wind-cheating burberries stowed on the front sledge. Not for long. In two hours, clouds spread from the west, and as they fought higher the cold intensified. They soon donned the outer clothing.

Looking backwards, in these first hours, they saw a great bare rock, standing free of the ice – an eminence that southern explorers called a nunatak. Mertz said it looked like a steep-sided island in a frozen sea; Mawson was tempted to retrace his steps to examine its rock structure, but realised it would be seen from the coast, that one of the two coastal parties would carry samples back to the hut – as he had instructed. He named it Madigan Nunatak, after the leader of his second coastal party.

Men and dogs pulled steadily upwards all morning, climbing onto frozen highlands, 2,600 feet above the sea. From here they could make out a deep swing of the coast to the south, forming an unknown bay beyond a peninsular of land east of Cape Denison. It meant a further detour southward before they could turn east. The surface degraded into wind-polished blue-green ice, the face of the great plateau canopy. At times they sped across the land as though it was an ice-rink. Then there came areas of compacted snow, windblown into solid drifts, so deeply frozen it was halfway to being ice. From these emerged the glassy, rolled tops of sastrugi on which the dogs slipped so that the sledges rocked and turned over, and Mawson and Ninnis suffered hard falls, with only their layers of clothing to cushion the blows.

Mertz had to discard his skis and found difficulty with the sastrugi, falling into the snow troughs when the distance between each ice peak stretched out to five feet apart. Mawson called him back to the lead sledge and himself took on the task of frontrunner-giving the dogs an object to pull for while ferreting out the easiest path forward. He had to maintain a kind of leaping run – five-feet strides – hopping from the top of one sastrugi to the next. He did this for a distance of two and a half miles and was surprised at the accomplishment. When they stopped to brew and make a longitude observation, he said to Ninnis: 'It was another cross-country run, Cherub! I must be much fitter than I thought.' Ninnis lifted an inquiring eyebrow. 'Why, D.I.? You've worked and trained as hard as any of us.' He had fallen into the role of leader, veteran; yet, he was still only thirty, and as strong as ever.

Soon after the tea break, they found their first mountain. They saw it loom through the snow murk some distance ahead, a shapeless black mass. It meant changing course to the east to meet it and excitedly they clambered along the sastrugi that ran in that direction. It was much further away than it seemed. The Antarctic light played tricks on their eyes and they marched for another five hours toward the dark rising mass before it sharpened in focus. They saw, then, a steep-sided mountain with a bare rock peak, lifting 2,000 feet from the field of smooth ice that licked its sides. Mawson thought he had found another extinct volcano. He guessed from the colour of the rock high above him that it was the same ancient, pre-Cambrian gneiss that marked Cape Denison... that it was a further clue to the antiquity of the lost land buried beneath the sea of ice. Thrusting from the frozen shield it would from now on be a known landmark for all travellers

in this place. He decided to name it after their ship –
Aurora Peak. Again he refrained from geologic inspection.
Frank Stillwell would be bound to see the black peak
and come to examine the structure. They built a cairn of
snow-blocks and he left a note saying he had been there,
that it had been named Aurora Peak. They pitched their
tent on the ice and slept well, almost in the shadow of
their first geographical discovery.

That morning brought a bone-cutting forty-mile wind,
which raised powdery drift across their path. Under this,
their feet slipped on a downward tilting decline. The
sledges took control as Ninnis fell, and went careering
down into the scattering, frightened dogs, who at once
were entangled in their traces. Mawson was anxious. In
the drift ahead, he found the ice floor canted sharply
down for 400 feet; in such a fall the sledges would have
smashed and the mission would have been aborted. They
tethered the dogs to ice-nails driven into the surface.
Then, they rigged a slipping tackle on an ice-axe, and,
with the three of them hauling back on the ropes, lowered
the three sledges to the next ledge, from there the ice
canted downwards again – to a total fall of 800 feet –
so the process was repeated. After three hours, sweating
under their burberries, they brought down the dogs and
made a noon camp.

In this arena of ice the wind passed mainly overhead.
Leaving his companions to recover, Mawson reconnoitred
their position. He wandered about for an hour, the sky

clearing above. He came back to the tent, eyes beaming with discovery. The depression in the ice, he said, was probably a crater: 'We are camped on the floor of an old volcano. The extinct crater is some 800-feet deep – and at least three miles in diameter.' The great ice sheet flowed over the formation, some hundreds of feet thick, hiding clues to some enormous eruption in an unknown past age. It was named, simply, The Crater.

They fought their way out of the sunken floor, scaling terraces of hard blue ice, slipping in banks of snow, hauling their loads almost vertically at times, and then, once again on the flat, reharnessing the dogs and setting off to cover more miles to the south-east, with Mertz ski-ing ahead on what first appeared to be a flat plain.

The late afternoon became a time of trauma. Two of the dogs were allowed to run free. Ginger Bitch, in late pregnancy and one of their best pulling animals, was in distress. Young Blizzard, the former hut pet and now a burly husky, had sustained a foot injury when struck by a falling sledge. Their freedom aroused feelings in the other dogs. Ginger (as distinct from Ginger Bitch) wriggled violently, escaped from her harness and raced back along the sledge tracks. Mertz turned and pursued her on skis, but she scrambled into the crater, out of sight. Mawson called his companion back. 'Let her go! If she comes back of her own accord, well and good. If not – I'd sooner lose a dog than have you break a leg down the ice slopes.'

They resumed their course – and within two hours they had need of all the strength they could command. Mawson was guiding the two connected sledges, controlling the dog team from behind in the low drift of snow blowing across their path – about a foot high over the surface. All at once they were in crisis; the front six dogs of his team disappeared from view; the snow bridge under their paws collapsed and they plunged, screaming with sudden fear, into a dark crevasse, writhing violently in their harnesses, and in their struggles dragging the other dogs and the sledges toward the edge of the abyss. With instant reaction Mawson dug his heels into the hard snow, gripping the dog traces with one hand, the rear sledge with the other. Suddenly he realised they were all on a single, wide snow bridge – covering a huge crevasse, at least twenty-five feet wide.

From the corner of his eye he could see that Ninnis was running forward to help with the pulling... 'Go back.' he yelled. 'Go back and get a rope!' If this whole snow bridge collapsed, a rope fastened into the ice might save them from final disaster... but, he couldn't risk waiting, the dogs were frantic, hanging over the edge of the broken area, and the strain was growing on his legs. He had to move backwards, slowly, easily, sliding one foot at a time, keeping his finnesko flat to the surface to avoid the 200-pound pressure of his body weight thrusting onto his toes, which might easily crack the surface. Foot by foot he reversed, his gloved hands aching with the weight of the dogs on their traces, with the heavy sledges dragging at his muscles; slowly his powerful frame won ground: and then, he stood on solid ice under a snowy surface. Ninnis was with him and they dragged all the dogs and the two sledges to safety.

He had no time to recover. A startled call from Mertz aroused them to the plight of the two loose dogs. Some distance ahead both animals were at the edge of a broken snow bridge, with their hindquarters below the surface and their front paws madly scrabbling at the rim. Mertz reached Blizzard; Mawson, skirting the broken crevasse, was just in time to catch Ginger Bitch by her hackles. He carried her in his arms back to the sledges and laid her on the ground. She looked at the three men and her eyes were soft and trusting, and then – as though in gratitude – she gave birth to the first of 14 pups; poor bitten, shrivelled mites, with no hope of survival.

Forced to back-track to avoid the danger of this riven area of the snow-covered ice, they again walked southward and faced a dipping sun, and the vivid light blazed beneath the overhang of snow clouds, bouncing from every shiny facet and glaring up from the sheets of white waste. It was late evening. Ninnis was taking a turn of front-runner, when, suddenly, he crumpled in pain, clasping his fur-gloved hands across his eyes. He was their first casualty of snow-blindness and its searing, gritty, pain. Mertz ran to his friend's aid, but Mawson called him to erect the tent as quickly as they could. Once inside the canvas cover, Mawson laid Ninnis on his sleeping-bag and applied the stock polar remedy – the insertion of small tabloids of cocaine and zinc sulphate under the eyelids, there to dissolve and ease the burning pain. Both of Ninnis' eyes were bandaged, to exclude all light; and when the hoosh was cooked and while Mertz spooned the hot mixture in his mouth, Mawson told Ninnis: 'You must wear your dark-glass goggles in that kind of light, Cherub. Especially when you're leading. I know you can't see well in drift with your breath misting them up; but, do remember, if you lift them from your eyes, screw your eyelids as tight as you can. You have to cut down on the amount of light reaching your eyeballs.'

Mertz went out to feed the restless dogs, and Mawson heard him burst into laughter. He put his head out through the tent entrance funnel, and there was Ginger back, sitting in line with the other dogs, eyes bright, tongue lolling in anticipation of her frozen dinner of seal meat.

As the famished animal crunched the meat, Mertz grinned at Mawson: 'Never shall we know, Doctor, why she ran away. But we do know her stomach brought her home.' They were happy to see her back; they were yet to know how precious she would be to them. As well, the numbers of dogs began to dwindle as they penetrated deeper into country of disturbed, pressurised ice. Hulky Blizzard broke his leg and he was helpless. Saddened, Ninnis elected to shoot him. They lost another three dogs in various ways, one vanishing, another falling down a crevasse, another too ill to do more than stagger.

That day set the pattern of delay and danger, which accentuated as they faced the sprawling glacier. This was nightmare country, a land of wayward, whirl-wind columns of sucked-up snow, with gigantic frozen cauldrons in the ice from which winds sweeping down the length of the glacier spewed drift high into the air, as though it was steam from a geyser. Here there were domes of ice, and open and snow-choked crevasses; here were frozen sink-holes sixty-feet deep, and steep undulations rolling into one another as the stupendous weight of the inland ice-cap pressured this frozen river to the sea. To the south there hung great ice-falls where the plateau canopy descended into the wide canyon the glacier had carved and moved downward toward the sea, spreading out to more than forty-five miles wide, feeding that floating tongue of ice, thousands of feet thick, which they had met aboard the *Aurora* last January. There were mountains rising far inland, and a coastal range through which this immense horizontal column of ice passed to reach the sea. Mawson saw it as land for geological wonderment. In those steep cliffs, where the rim of the continent's colossal frozen mantle spilled over into bluffs and motionless ice-sheets and ice-falls, were hidden strata of rock, probably as old as any on the planet; ancient landmass that, if his clues were a guide, might hold ores of precious minerals, and coal and oil-bearing shales... As they moved from the banks of the glacier, to fight their way through a labyrinth of winding crevasses, onto the bed of the greatest glacier then known, it became apparent that those inland mountains were guarded by barriers immensely forbidding to mining.

–

The days of battle with the glacier dragged at their time and energies. They fought for six hours one morning, starting at 7 am, and trudged five miles to make a forward advance of only two. Mawson slipped into a crevasse, to the armpits, and was pulled out by the rope attached to the sledge from his own harness. While marching, now, they roped themselves to the sledges as precaution. On the third successive day they halted for lunch among wildly wandering cracks in the surface, and Mawson and Ninnis walked a few yards to photograph a widemouthed crevasse that seemed to sink to bed-rock. Returning to the camp, they passed either side of the tent. Mawson heard the cry of alarm and a clank. On the other side of their pitched tent, Ninnis was in a crevasse, being held up by gripping the legs of the camera tripod. Mawson hauled him out and they stood together in a moment of fascinated fright, looking down into a sheer-sided pit of ice, too deep to estimate. That night Mawson pencilled a note into his diary: 'We found the crevasse to be fifteen-feet wide and very deep. Ninnis had a very narrow escape. When we looked again we found we had actually pitched our tent on the snow bridge over this chasm. Needless to say, we lost no time in breaking that camp and moving away.'

To reach their camp that night they faced a great field of ice – within the glacier – where the pressure had pushed the surface into wave formation as though it was an ocean frozen instantly at the height of a storm with wild peaks and pinnacles of shattered, jagged ice. Again and again they turned back on their tracks, edging forward a few miles but walking, climbing, hauling down slopes, as many as sixteen miles a day. The eastern side of the glacier became a citadel to them. Slowly, remorselessly, the continent was wearing them down.

They came on further valleys within the glacier bed – valleys of ice up to 200 feet deep and a third of a mile across, valleys that were mere ripples in this frozen river flowing to the ocean. In between, there were sudden crises. In a day of drift, after a night of seventy-mile wind that shook and rattled their tent so their sleep was broken and left their awareness blunted, they came close to the ending.

The snowflakes that morning were huge – six-pointed frozen stars that mingled with the blown drift. They had stopped for Mertz to find the next half-mile of passable terrain, and Mawson stood by his dog team, waiting, with his hand on the main tow-rope between the dogs and the two joined sledges. All at once, in that eerie light of floating snow, he was aware of a strange sensation. His feet were not moving, he was making no effort, yet his body was gliding forward. Then he realised. He was not going forward; the only objects he could see and associate were his dogs and the joined sledges, and *they* were moving backward. As he stood there, waiting for the call from Mertz, the rear sledge, silently, stealthily, had sunk into the snow bridge of a crevasse. Its gliding weight was dragging the front sledge backwards, was dragging the dogs backward – to disaster. He yelled the call for help with all the power of his lungs as he leaned his whole weight in a frightening tug-of-war with the dangling rear sledge. Gradually, he lost ground; inch by inch his heels were sliding through the snow. Then, thankfully, Ninnis was there, hammering an ice-axe into the surface, and Mertz was bending a rope around it to take the weight from his straining back.

It was one of many struggles and exertions that the glacier imposed. They hitched all three sledges together for a trial, but the last, and weightiest, sledge crashed a crevasse twice, and each time Mawson was lowered on a rope for hours of carefully lifting each seventy-pound tank of dog-food for hauling to the surface, hanging and straining in a situation where a false, or hurried, move would have sent the whole sledge load crashing to a bottomless pit.

Mawson was so drained, he fell into his bag, and forgot to wind his timepiece. The watch stopped in the night, and the aid of knowing the exact time at the longitude of the hut was lost. He followed meticulously the daily routine of making positional observations, latitude and longitude, of noting meteorological aspects and all points that would aid his map-making plans on which he would base his claiming of this new land for his country. He was bound now to fall into some inaccuracy – and that offended his scientist's mind, and added another worry to the nagging stresses of this difficult mission.

Now, they were near the eastern side of the glacier, and, gratefully, they saw the field of bright ice rising hundreds of feet beyond. There were more hours of labour, manhauling the sledges up the shining slopes, then they were standing high above the torturing glacier and could look back over the land they had crossed.

The glacier lay like a frozen serpent across the land-scape, twisting one way then the other, and beyond, dark against the sky, they could see Aurora Peak. Mawson jabbed his fur-gloved hand in its direction. 'There, gentlemen,' he said, 'is a beacon that will lead us back to the hut.'

They turned their backs on the glacier and trudged inland to go more directly south than had been expected. They had been in the field for two weeks and had advanced 120 miles – a quarter of the distance Mawson had hoped for. Now they faced weeks of marching toward a point some 380 miles east, and there were already doubts in all their minds on how far they could go before they were compelled to turn back. The continent was sowing the seeds of depression.

For two days they slogged across the ice of the upper plateau, 2,500 feet above sea level. They were buffeted by forty-mile winds, which, with the temperature well below freezing, chilled them to the bones. Plugging the miles away over difficult and changing surfaces, Mertz breaking trail when there was snow, ringing the changes between the sledges, they came to where the frozen plateau tilted into a downward slope, and, as they slid and skated, hauling, dispensing with the dogs, and struggling through drift, the wind fell away and they saw a splendid scene rolling at their feet... another immense glacier winding crookedly out to sea, and going back deep into the untrodden inland... a glacier much bigger, more impressive than the one they had left behind – its awkward, floating tongue so far out into the ocean, it topped their horizon. It was yet another daunting, fearful barrier for them to surmount.

On the far side of its mouth there stood, some sixty miles away, a huge colonnade of rock, sheer from sea level, topped by the ice-cap. These cliffs were at least 1,000 feet high above the shore, with vertical formations – like organ pipes of dark, black rock – extending at least sixty miles along the east coast. Mawson named the structure Horn Bluff and guessed – correctly – that these were eroded columns of dark dolerite overlying a coal-bearing sedimentary formation. It was a geologic magnet, which he resisted, knowing that Madigan's group would be certain to fight their way this far and make a detailed inspection. For his party the way led south-east, curving away from the southward turning coast, and, immediately ahead on that route, was the crossing of this vast glacier.

–

They spent a whole week in its fastness. They discovered a great ice-coated island in the glacier stream; they battled with the expected upheavals and with tortuously winding crevasses, and lay anxiously in their tent under sixty-mile winds with the poles and canvas shaking, flapping and quaking above their heads. Men and dogs fined down; all were ravenous for more food. There was snarling among the dogs with a hierarchy established on the basis of brute savagery. The dominant male was Basilisk – named for his ferocious snarl and evil eye – with his main challenger Shackleton, not so hefty, but, sharp of fang and very agile and quick… so quick, that one evening when, in a moment of carelessness, Mertz left him enough trace to reach a sledge, he ripped open a food bag with his sharp teeth and gulped a two-and-a-half-pound pack of their energy-giving butter. They would eat slices of it for their lunch; they needed knobs in their hoosh to give it added nourishment; it was to be sadly missed.

–

They emerged from the glacier at the end of November. The worst ordeal of their outward journey so far was behind them; they looked ahead to a rising ice slope with relief. Mawson said solemnly to his two companions: 'I vow never to lead you again into such terrible ice. We shall do everything possible to avoid a repeat of that experience.'

He was acutely aware that the conditions had made deep inroads into their morale. When the sun came to light their scene, it, too, turned into a weapon to wear them down. The way upward from the great glacier was scouted by Mertz on skis and they struggled onto a level surface 1,400 feet high, to find themselves soon among sharp-edged sastrugi – which made it impossible to use the dogs – with small fields of soft, melting snow in which they and the animals often were bogged, wading up to their hips in soggy drift.

Mawson wrote: 'We have had an aggravating morning; very deep, soft snow and climbing hills with plenty of crevasses. The dogs very done. Things are looking serious for our onward progress. If *only* we could have a straight-out sledging proposition instead of these endless snowhills and crevasses.'

On 1 December, the temperature rose above freezing for the first time. It was horror. Toiling their way to an upland they went through 'sticky snow with the consistency of thick porridge.' He noted: 'We plugged on uphill for several miles only by the greatest exertion… the sledges were hard to move, even with the men and all the dogs pulling their hardest. We now have to consider travel at night when the temperature is lower.'

They were high above the second glacier, and again their eastward thrust was diverted by open water. Where the Wilkes expedition had reported land, and named Point Emmons and Cape Hudson from ship-borne observation, there was nothing but sea, dotted with ice-floes and lonely, distant bergs.

Indeed, the mass on which they stood was solid ice, a projection of the canopy overflow; there was no land underfoot, no land immediately east, and again they would turn south to skirt an indentation to go toward that point west of Cape Adare. They could march till late in the day. At midnight the sun still peeped over the south polar plateau, licking the ice into frozen gold.

On the top of the plateau they came on strange ice formations, flat-topped sastrugi by the thousands, blocking their path, all about three feet high and shaped like anvils. In that country the plateau blizzards came back to torment them. They fought a fifty-mile wind to get their tent up, built a snow wall to shelter the dogs, and then spent a restless night with the gale reaching above seventy-mile strength and shaking the poles, rattling and flapping the cover.

In the early hours, Mawson became aware that Ninnis was sitting up in his bag. He held a book in his left hand but his eyes were shut; his right fingers were curled around the bowl of his calabash pipe, as though for comfort. He was rocking backwards and forwards, as though he might be soothing a baby. Immediately, Mawson remembered Madigan, in his sleep, searching the tent for a pricker to unblock the jet of the primus stove...so, he gently pressed Ninnis back and covered him with the cowl of the bag.

In the morning the gale still blew. Mertz went to find the dogs under their blanket of snow and feed them; and so thick was the drift he could not see the tent from three yards distance. They had no hope of moving on. So, Mawson cut down their daily rations from thirty-four ounces of dried food to fourteen, and they rested in their bags, all that day, all that next night. Again, Mawson found Ninnis sitting up hugging his pipe to his chest. At another time in the night, he started writhing and shouting commands to the dog team of his dreams, calling into the white whirling night: 'Hike! Hike!' He dozed during the days of the blizzard, and, at other times, read to them from the Thackeray book his mother had given him... and Mawson could detect a flagging note in his voice.

To help pass the time, and to cheer them, Mawson asked them about their homes and their lives. Mertz regaled them with stories of the social structure of Basel, of the pride of the old families and the rigid traditions, of his father as a wealthy manufacturer of air-conditioning plant, and of his stiff-backed mother and his four brothers and sister. Ninnis talked of college in Dulwich, of his doctor father and the family's history in the Cornish tin mines, which went back to biblical times. Mawson added his recollections of his time with Shackleton, of his meetings with Scott, of journeys into the Australian outback, of the first uranium field to be found in Australia, at Mount Painter in northern South Australia. They talked of the men at the hut, conjecturing on the progress of the Magnetic Pole party... then Mertz recited passages from Sherlock Holmes, and they nibbled at hard biscuit, and slept, always restless and feeling the ironclad face of the ice through their reindeer skin.

Four nights and three days the blizzard raged. Then, at 5 am on 9 December, they were startled by the sudden silence. The snow had banked up two thirds of the way on the outside of the tent, pressing in the covering, and they had to dig their way out of the funnel to a world so still and white, it was eerie. It took nearly three hours to break camp, digging out the dogs and the sledges, preparing food, thawing their stiff garments, harnessing and packing for the trail. At 8.30 they turned east, threading among the ice anvils, now all surmounted by mounds of snow. The sun became visible through a white, hazy curtain. It was a cold, liquid, shining globe with a brilliant halo.

The snow was a newly laid carpet and skis and sledges glided across the surface. It was then nine degrees below, but, soon, the sun pierced the haze and the softening snow became purgatory for men and dogs. After six hours marching Mawson called a halt for lunch. Ninnis had been lagging and looked tired, despite three days in his bag.

He could give the young man no respite, but could leave him in last place so that all he had to do was follow. It was approaching a crucial time for their dash into the unknown eastland and Ninnis' apathy was worrying. They put up the tent to make tea and carry out the altitude test, which needed shelter from the wind and boiling water. They had come down about 500 feet and were then 2,300 above sea level.

They went on, marching till midnight. The toil showed on their faces and their actions. Mertz was cook and his hands trembled as he struck matches to coax the alcohol to burn to ignite the primus flame… and he knocked the bottle of alcohol against the stove and broke the top, spilling half the spirit into the groundsheet. Mawson looked up, concerned, from his notes, and Mertz had anguish in his face, tears were forming in his eyes. The awareness of what this land was doing to himself and his comrades was blindingly sharp. Jolly, goodhearted Xavier, shaking, almost in tears over spilt spirit; quiet, loyal, young Cherub losing zest, growing apathetic, and his own condition deteriorating – he had a swollen underlip, painful gums, and the left side of his face ached all the time with neuralgia. He knew he would suffer more than the others; the cause and effect was obvious. He was bigger than both of them with a larger skin area; heat loss was greater, and heat was body energy generated by food… and they meticulously all had the same amount of food. He knew the crushing strain on Ninnis from his memory of the 1909 march; he knew the drag on the muscles and the burden on the mind. He had to put a limit on the pressure, to cut his own ambition to their condition, their spirit, their capability.

He fell asleep that night, his mind working on the miles, the food, the days they had left before the *Aurora* was due back in Commonwealth Bay. Even that desolate wind-bitten rocky shore now seemed a most desirable haven.

–

In two days of solid marching, they covered a further thirty miles across the plateau. On the second night, noticing Ninnis favouring his right hand, he saw the second finger. At the top, around the nail, it was black and green. The finger was grossly swollen, bloated by pus built-up under the skin to twice the normal size. Mawson asked to examine Ninnis' hand. He was shocked; he could almost see the pounding blood, the throbbing that had been denying Ninnis his sleep. 'It's a wicked whitlow, Cherub. How long?' Ninnis lowered his eyes. 'Little sleep for a week now, D.I. I'd hoped not to bother you, that it would burst – or just get better.'

After hot hoosh and cocoa they tried to reduce the heat in the finger with a snow poultice. Mawson knew it was time to lift all their spirits, to talk about turning back to the hut. They had made three good marches on their last stretches. If they could do two more solid days, and then – on 14 December in the evening – unload the food they would need to get back to Cape Denison and leave it in the snow, they could make several forced marches with a lightweight sledge and two dog teams. 'If we can do that, if we get fair weather – and I think it'll get better as we go further east – then we might get very close to the 155 east meridian, about where I was with David and MacKay in 1909.'

Mertz was serious, concerned for Ninnis. 'How far – how many miles we walk?' Mawson said he calculated two more days of good marching would take them 340 miles from the hut, and the race into the east would mean a total distance out of more than 400 miles.

Mertz said: 'And then – we have to get back, no?' They would try to make their return journey inland, to see more of the mountains and try to avoid the terrible crevasses in the two great glaciers. 'But – two more days, and we unload the weight. Then, two or three more days and our faces will be looking westward. That's good to know, isn't it?'

It was 1 am, in the morning of 13 December. As a last act, Mawson laced some warm water with a teaspoonful of alcohol and gave it to Ninnis to help him sleep.

–

In the morning Ninnis confronted Mawson. His once pink, cherubic face was grey and cracked with wind and cold, his eyes were bloodshot and outlined with the mauve of sleeplessness.

'It's no use,' he said. 'The pain is simply unbearable. Would you cut the finger open, please?' Mawson hesitated. The thought of an open wound on the hand in these conditions of forced marching in wind and extreme cold gave him pause. But, then, this pus was bacterial poison that might enter the bloodstream; septicaemia was a greater peril. 'Of course, Cherub,' he replied. Then to be reassuring: 'It'll hurt, but if we can relieve the pressure it should reduce the pain. You'll go to the bag again for a few hours after, and Xavier and I can do some re-packing.'

The operation was swift and simple. Mawson sterilised his knife in the primus flame; Mertz cuddled his friend's head into his stomach, with an arm across his eyes. Mawson chilled the swelling with ice, then quickly ran the edge of the blade along the side of the finger, squeezing as he did so. The green-yellow pus spurted, then welled, and dripped into the snow. Mawson smothered the wound with a pad soaked with iodine, and Ninnis gasped with the burning sting. It was done. More water laced with alcohol, and he crawled into his sleeping-bag.

While Ninnis slept, Mawson and Mertz revised their sledges. The one that Ninnis had handled was now well worn, having carried the heaviest weight, and was also damaged from its tumbles into crevasses; it was decided to discard it. The food loads were stowed on the best of the two other sledges and this would be the new, rear sledge. It would carry their tent and the main essentials, their reasoning being that their second sledge would always cross snow bridges safely where they had first been negotiated by the front-runner and the first sledge. When the repacking was completed, the rear sledge carried about half-a-hundredweight more, but Mawson reckoned that a few days of feeding men and dogs would soon even out the two loads. They also discarded a broken spade, the lighter one that had proved inadequate for rapid cutting of snow-blocks; they threw away old clothing, worn mitts, tattered finnesko and broken straps. They would travel faster now with two teams pulling two sledges instead of three. And the dogs, too, had to be reassembled. There were twelve dogs left. The best were the two ruling males, Basilisk, Shackleton, with Ginger Bitch the true, loyal puller, tough old John Bull, Castor, and the errant Franklin. These half dozen were the strongest; they were the dogs who would pull the heaviest sledge and march with the men back to the hut. The other six were lazy George, Mary, Johnston, Haldane, runaway Ginger and Pavlova – who was Mawson's favourite. When the seal meat and biscuits ran out, when hunger made the animals wolflike, these six already inferior, skinny dogs would be shot and fed to their mates… that way, at least six animals would return to the hut.

Ninnis woke refreshed. His finger was very sore inside his glove, but it no longer throbbed maddeningly. Mawson spoke to him, kindly, relieving him of any duty but riding or walking with his dog team, in the rear.

Over the unknown, thrilling country their convoy wended, again turning inland to avoid cascading icefalls, skirting the coasting ice where it had split into wide chasms. Then, over the brow of the canopy came a thick, cloying fog, closing around them, blanketing their vision like billowing smoke, and smothering sound as well as sight. The fog brought Ninnis a strange, lightheaded sensation. He was not here, walking an unknown land at the bottom of the Earth: this was the day he'd felt his way through the foggy London streets, through the slush in

Regent Street, to Mawson's rooms, there to stand smartly and salute and say, 'Thank you. Thank you very much, sir', when told he'd been selected to go to the Antarctic – all the way by ship, taking care of the wonderful dogs. Out here in the fog he could imagine the tall buildings on either side of him – only yards away – and he was walking back along Regent Street, thrilled, proud, elated.

They left the fog behind. Ninnis could see they were treading a long, vanishing incline with the ice waves between one mile and four miles apart and the snow troughs 500-feet deep; high in the sky ahead, miraged above the white horizon, was a seascape of myriad islands, and bergs snared in a frozen ocean. The snow here was near to being ice, crisp and hard underfoot. Soon, as they moved further down, Ninnis heard the sound of protest at their presence.

From behind there came to his ears a deep, thunderous rumbling – the boom! boom! of a thousand cannon, firing in unison far away. Mawson said he thought they had walked over a frozen crust, and that their weight had caused cracks which spread into the frozen snow and released air trapped under the pressure of the ice. Boom! boom! – the sound followed them as they moved east and roused apprehension in Mawson's mind, the ice was sounding a warning. They camped that night among formations that obviously flowed to feed a coastal valley.

Mawson wrote his presentiment into his diary: 'We have a long downhill run… we are apparently coming to another great glacier, with all its attendant problems.'

Mertz and Mawson performed the chores; Ninnis' right hand was in a sling and it needed rest for speedy recovery. The tent was warm and cosy with the blue flame under the hoosh pot, with hot cocoa and biscuit and a stick of chocolate to follow. Ninnis rolled up in his bag and Mertz tied the toggles under his chin. Mawson patted his shoulder as they lay side by side. 'Have a good night, Cherub.'

In the early hours a forty-mile wind blew streams of snow over the tent, rattling the poles for four hours. Ninnis knew nothing of the storm. He woke at nine to a lovely day, a half-clear sky with broken cumulus letting the sun come through, and a ten-mile breeze. He delighted in the thought he had had a good sleep.

It took two hours to break camp. Mertz and Mawson did all the work. They fed the dogs, loaded the rear sledge with tent and poles, tools and instruments stowed with the food tanks and seal meat and dog biscuits. Before they started, Mawson told him: 'Take it easy today, Cherub; give that hand a chance to knit. The surface looks good and we'll be downhill and you'll be able to ride the sledge best part of the way.'

It was a sleigh-ride for the first few hundred yards. Ninnis saw Mawson look at his watch after they'd gone a quarter of a mile and wave for a halt. It was time to make the noon sighting of the sun. They indulged themselves with a snack and a boil-up of tea.

It was an hour later when they set off again. Xavier Mertz was breaking trail, about thirty yards ahead of Mawson, and Ninnis could hear his singing voice.

Mawson was sitting on his sledge, writing in his journal. Ninnis sat side-saddle on his. The dogs were pulling well down a gentle slope.

In front, the sound of singing stopped. Ninnis looked up to see Mertz holding a ski-stick in the air pointing to the right. It wasn't unusual. It indicated he had skied across a crevasse that wound away to the right. Mawson saw the signal; he looked up briefly, half-turned his head toward Ninnis, and shouted: 'Crevasse!' Then he bent again to his notes.

Ninnis knew exactly what to do. To cut short his time on the snow bridge he would move to the left, go directly across. He could redirect the dogs better on the ground rather than from the sledge, with his one good hand tugging at the traces.

He jumped from the rear of the sledge. His toes hit the surface like a heavily weighted dart. His body plunged through the disintegrating snow; and the sledge was plummeting, tumbling with him. He heard the frightened wail of his lead dog, Basilisk, as the impetus of the sledge dragged all the dogs into the depths. Ninnis opened his mouth to scream his terror. The fine snow choked his eyes, ears and throat, and he did not hear his own smothered death cry. Down in cold blackness, 150 feet down, his falling body smashed into a projecting ledge of ironclad ice. With the shattered remains of his sledge, with the doomed dogs, Belgrave Ninnis plunged deeper, and deeper, into the abyss – into infinity.

SIX

In Peril on the Ice

The shock of the death of Ninnis did not hit them instantly; even when they gazed into the awful hole they could not grasp he was dead, and the peril in which the calamity placed them was slow to dawn.

Mawson had squatted at the back of his load busy with his noontide notes; he'd seen the ridges of frozen snow – which Mertz had signalled – glide obliquely under his sledge runners. He saw nothing menacing in their appearance; they had crossed hundreds of such ridges in their journeying. Yet – out of precaution – he called the warning to Ninnis, looking over his shoulder as he did so and seeing the young man, right arm in a sling, rising to straighten his team for a direct crossing. Moments later he heard the only clue to the tragedy; Basilisk, was whining in protest. It seemed so obvious he was being chastised for not answering a command that Mawson called to his own laziest dog: 'And you'll get a touch of the whip George, too, unless you pull better!'

Mawson had much to ponder just then. The noon reading had put them within striking distance of his objective, inland from that location where, aboard the

Aurora last December, the unyielding pack-ice had denied him a landing for his main base party.

Tonight – when they would be about due south of the city of Sydney – they would ready the sledge for starting the eastward dash in the morning. There was little to do, other than re-lash the two dog traces so that both teams from tomorrow could pull a lightened load. His leading sledge already had their week's food rations aboard. They need only dump the bulk of the kerosene to save weight, take on a week's dog-food and the tent and tools and groundsheet; then they could leave Ninnis' sledge in the snow to be picked up on the return; for three days or so, depending on weather, with a light single sledge, Mertz on skis, they could literally race over the miles – if the surface was good. Would they get near enough to see those great mountains he knew ran into the sky west of Cape Adare? He liked to mull on the thought – target achieved, and land found where no other men had walked... and it was a grand afternoon to plan this last dash. He wrote: 'As fine a day as any we've had on this trip. Temperature 21 F; wind a ten-mile breeze from ESE; the altitude at noon...'

He was aware that Mertz had halted in his tracks ahead. He was staring back along the trail – and some element in his stance roused alarm. Mawson stopped his dogs and looked over his shoulder. The single sledge track ran back for a quarter-mile. From there two tracks went beyond to meet a hazed horizon. And the landscape was bare – empty. Afraid, he ran hurriedly along the track. Soon, he heard the thin sound of a wailing dog and there was the hope that Cherub and his team were wedged in a crevasse again, hanging in their harness, roped to the sledge, waiting for rescue.

The gaping hole in the frozen snow bridge killed the hope. From near the edge of the dangerously cracked rim he peered into an icy pit and an awful feeling of catastrophe gripped him. He turned and waved frantically to Mertz, shouting to him to bring the team and the ropes, and again he heard the wail of a dog, coming out of the depths, a mingling cry of pain and fear.

Mertz reached the hole and at once was almost distraught. Appalled, they both circled the crevasse to lie down on the edge of solid ice, to call and call into the depths. Only the dog answered. Mawson found his binoculars, Mertz held him by a rope tied to his waist and he leaned over, peered down... the hard sheer walls fell away from green and blue ice into deep darkness, beyond light, into the black, cold heart of the glacial depths. Down in the murky half-light he could see a projecting ragged ledge of ice; on the uneven surface was a wailing dog. It was Franklin. His spine was broken; he struggled to stand up and cried with the pain and frustration. There was the body of another dog on the ledge and shattered remains of Ninnis' sledge; there was no other sign or sound of life. And very soon Franklin died. The two men shouted Ninnis' name into the crevasse, their voices echoing and resounding from the ice walls.

A steady draught of intensely cold air came up to them.

It was all beyond their reach and repair. Mawson forced his mind to accept their helplessness – Mertz would not.

Frenetic in his grief, he wanted to try to lodge the remaining sledge across the lips of the crevasse, to climb down on what ropes they had. Mawson had to restrain him from recklessness. The hole was too wide for the sledge, the ledge too deep for their rope. To find how far down it was to the ledge, he produced his fishing-line, 150-feet long, and tied the legs of the theodolite on the end and lowered this into the hole. The legs just touched the back of the dead dog.

Yet, for three grief-stricken hours they circled the broken snow. Then Mawson faced the savage fact. An injured man, flung into the terrible cold of the ice-depths, would not recover consciousness after three hours – nor could he be recovered. This was their companion's grave; now he had to accept his leadership, his responsibility to the living. In a gesture of comfort, he threw an arm round Mertz's slumped shoulders and led him away to the sledge.

For Mawson, all the hopes and dreams lay shattered with the fragmented sledge and Ninnis' broken body at the bottom of that hole. He told Mertz that the exact position of their companion's grave would have to be marked on their chart of this new land. They would know the spot, thereafter, as the Black Crevasse... but they had to be certain of its location. They would need, he suggested, to reach a high point in the ice plateau to be sure of the position. And so they climbed several miles, onto the high ice, nearly 3,000 feet above the far sea, and saw the vast panorama of white waste running eastward, beyond sight – land they would now never walk together. There was no hint of distant mountains, only endless, coastal ice-cliffs, blank-faced, high, white, with no sign of rocky shores.

Northward from their perch the sea was mainly open, glinting between icebergs and the islands that Ninnis had sighted two days before. It showed clear around clusters – floes – and that meant there was no prospect of fast travel back to the hut over a frozen sea.

Now the daunting facts of their own plight took clear shape in his mind. They started back to the sledge, and Mertz ran ahead to the hole. Once more he was calling Ninnis' name. Mawson tried to comfort him again and Mertz turned, face twisted in anguish: 'I know it is the hand of God; but why? Why?' With the question he lifted his eyes to the sky... and, there flying above their heads, circling on beautiful white wings, was a silver-grey Antarctic petrel. Mertz watched it, silent, entranced, as it hovered a moment over the dark crevasse; then, with a tilt of its wing-tips, it swooped across the snow, over their heads and into the hazy north; this was the first wildlife he had seen on this journey. Now his emotions travelled too fast for his command of English. 'He was *mein besten freunde in der ganzen* expedition, Herr Doctor. The wild petrel could be his soul, no? Ist a legend, ja? The sea-bird is a dead sailor... perhaps this petrel is Cherub, free for ever.' He had come to accept the truth. Ninnis was dead. Now Mawson drew him to the present.

'We are in dire peril, Xavier. We must together consider our position and what we must do.'

Left aboard his sledge was a single food-bag, sledging rations for three men for a week – pemmican, sugar, dried milk, a pack of butter, a tin of cocoa, two dozen biscuits. In his own personal bag was a small box of sticks of dark chocolate and a bag of raisins. All reserves of food, all the dog food, was at the bottom of the icy abyss, along with their heavy-weather tent and the poles, with their ground cover, the spade and the pick-axe, the mast and spar for erecting a sail in suitable weather, their mugs, plates and spoons. Mertz had also lost his waterproof burberry pants and helmet. The six strongest dogs they'd hoped would help pull the load back to Cape Denison were now dead.

'We are some 320 miles from the hut and we have been out now for five weeks. We have the barest resources to get us back, to face all the dangers; so – you know what we have to do to stay alive?' Mertz knew and understood. His eyes were dark and solemn. He nodded. 'Yes,' he said. 'We shall have to eat the dogs.'

In this dire plight the broken sledge and the old worn materials and cans they had thrown away at the camp of the night of 12 December were suddenly precious to them. A dump, fourteen miles back along the track, had to be reached for even a rudimentary shelter. The lightweight drill tent, used until now to cover Mawson's sledge-load, was their only protection against blizzard and wind and cold, and it needed supports that could be cut from the discarded sledge. It would mean slogging back across the ice, non-stop, for many hours, but the effort would help to ease their grief, Mawson felt.

Before leaving they paid their last respects to Ninnis. At the rim of the crevasse Mawson read the burial service from his battered prayerbook: 'The Lord giveth and the

Lord taketh away...' They stood bareheaded, their breath white clouds round their faces. Mawson spoke his eulogy to a lost, loved companion, a fine selfless person and a brave young soldier: when men faced lonely blizzard-swept wastes together, he said, when they shared toil and want, and suffered hunger, cold, weariness, they were ever after bound to each other in some wonderful fabric of fellowship and affection. They would honour their lost comrade by naming the second glacier they had discovered, the greatest glacier then known in the world, the Ninnis Glacier. Thus, while his grave would be eventually lost in the shifting ice, his name would remain on the maps of the world.

Dr Xavier Mertz, wet-eyed, spoke his own eulogy: '*Freunde Ninnis, Lieber alter Ninnis ist tot...* he is dead, and we must try to live for him.' Deeply touched, he wrung Mawson's hand: '*Danke, danke!*' he said.

With the ski-shoes they rigged a small shelter over the primus stove. Mawson found two old food-bags, still dusty with the pemmican and milk powder they had contained. He ignited the primus after some trouble, and, when the snow had melted to water and was steaming, he put the two bags into the hot liquid and left them to boil for ten minutes. Mertz readied the dogs, tightening their harness.

They sat in the little lean-to and supped the cloudy, hot liquid from each side of the can, feeling the first warmth in their bodies for nine hours. The dogs were ravenous. They lifted their muzzles to the sky and joined in long, mournful wolf howls – a pitiful, aching, animal call of hunger sounding over the frozen wilderness. Mawson went to them and patted their heads – he had nothing to give them; all their sustenance had vanished into the maw of the Black Crevasse.

Eight hours after Ninnis was killed – at nine in the evening of 14 December – they pulled with the dogs along the old tracks, until they came to the rise that would hide Black Crevasse from their view. They halted briefly, to look back for the last, silent farewell. Now opened the long, painful march to the west.

They did not stop again that night. They were driven on by the crucial urge to reach the previous camp. Where the snowy slopes rose they plodded doggedly on, where the line ran downhill Mertz – on skis – and Mawson on the sledge allowed their load to rush downwards, without regard to danger in the crevassed slopes, knowing the risks were there with all the consequences, but disregarding them with a languid carelessness, which Mawson knew was caused by the death of Ninnis. Through the hours of the hovering midnight sun, onward they marched into the early morning light, a light blown snow filling their goggles and raising their silent prayers that there had been no snowfall to mask the location of their precious rubbish dump.

The blur of the dark smudge came through the white haze at 2.30 in the morning; dogs and men, with an objective ahead, rushed towards the site; then, with wearied, wobbling legs, trembling from the long-sustained effort and from cold, they all dropped into the snow. The animals lay with their heads on their front paws, crying from hunger and tiredness, their eyes sad and their breath forming crystals on their faces like frozen tears. Mawson regarded them with utter pity: old George, defeated, unable to walk again; Mary, and Haldane, and the old battler Johnson – who'd stood up to a three-ton sea elephant – then gallant, loyal Ginger... and on the end of the line, dear Pavlova.

Something had to be found for them this night. He rummaged among the discarded materials and found two holed wolfskin gloves, a pair of worn finnesko boots, a short length of hide strapping. With his sharp sheath-knife he cut them into six equal portions; then he and Mertz fed them to the dogs, one by one. The prospect of something to swallow was electrifying to their wild hunger; yelping, howling, they each gulped their portions in a flash. They licked the snow for the last trace, for the last hair. There was no more, and they buried themselves into the snow to sleep as well as hunger would allow.

The two men needed shelter, and, with Mawson's Bonzer knife – an adjustable complex to which they could attach a small saw, a little hammer and a file – they cut a runner from the old sledge. Mertz sawed this into halves and then, with frozen fingers, they had to lash the struts of wood to the two ski-shoes, to fashion a frame for their makeshift tent – a nightly task that was to bear increasingly on them in the weeks ahead. At last they threw the light cover over the squared frame – praying there would be no strong wind – and pinned the skirt down with snow heaped by hand and with pieces of their sledging equipment. There was room only for two sleeping-bags rolled out on the floor of bare snow; the peak of the crude little tent stood a mere four feet high so that only one man could sit up at the one time. Yet, this was blessed shelter without which they would be frozen corpses in the snow within a day or two.

The primus was lit, its blue light and its roar a comfort to them after this terrible day; they did not eat, agreeing that their food would be saved until they were able to enjoy it to the full. The snow melted in the cannister, Mertz added a few drops of alcohol and they eased their thirst and felt the warmth of the spirit with gratitude. They were still in shock, and worn from the long miles of tramping, yet both wrote their diaries, lying face down in the snow in the tent.

Mertz wrote: 'Mawson and I have a long way to go and we must stick together and be good comrades to reach winter quarters. God's ways are hard to understand, but I must try to serve Him as best I can. It is a consolation now to my mind that Cherub died so quickly. Now he can rest in peace, the dear friend.'

Mawson located the glacial grave with some precision: 'It is thirty-five miles east-south-east of the headland which I named Cape Freshfield; fifteen miles from this charted camp.' And his mind was on the dogs, already near starvation and complete exhaustion. 'Winter quarters are 300 miles away and we shall not get through unless we eat the dogs. We are terribly handicapped.' He pondered their plight and the fearful struggle that lay ahead of them. He added a further separate line and underscored it: 'May God help us.'

The new day burst on them six hours later with the noise of snarling, barking and gnawing teeth. The two men dosed uneasily, discussing the best course to the hut. Then, in mid-morning, when the dogs were barking wildly, Mawson thrust through the tent funnel, aching and dispirited, to find Haldane and Johnson – at the full stretch of their tethers – gnawing the wood of the old sledge; a leather strap that had been within their reach had been eaten half away. Three of the other dogs were straining and leaping for a share. Poor George lay apathetic, too weak to rise on his legs. He would have to be the first to die, to feed men and dogs.

Mertz could not face it. Mawson took the .22 rifle from its place on the sledge and led George behind the tent – then, saddened, shot him through the brain. The sharp crack of the rifle stilled the five tethered dogs into startled silence – for a moment. When Mawson dragged the body for butchering on the old sledge, they were already slavering in expectation.

Mawson cut away the fullest leg muscles and then excised the liver, surprised to find it had not diminished in size due to the malnutrition. With Mertz, now sawing, they cut up half the carcase and fed it, with the offal and the head, to the five frantic animals. Their fangs ripped sinew from the skeleton, and tore tail, offal, pelt into pieces that could be gulped, then five pairs of ravaging jaws cracked and crushed the bone, and tongues licked away the marrow. In minutes, nothing of George was left in the snow. They had much to do before they would eat their share. Mertz cut a strut from the old sledge, and, with the knife, patiently shaped two wooden spoons. Then he repaired the vital spade, its broken handle was spliced with other strips of wood and bound with lamp wick. Mawson shaped two disused tins into food and drinking pannikins, and then sewed up tears in the tent cover.

–

They ate their breakfast in the afternoon. Mawson decided the snow was too soft for travelling by day and they would make more miles tramping over the surface when it froze at night. First, they carefully apportioned rations from their dried food bag. Mawson observed: 'We have been five weeks coming this far; it will take the two of us at least that time to get back – with the best of luck. There are roughly ten days of normal sledging rations and so we must reduce our normal intake of thirty-four ounces a day down to eight, and pray that the dogs give us enough nourishment to keep marching at a rate fast enough to get us through.'

He used the lid of the Nansen cooker as a frypan. George's two hind-leg muscles were fried, but there was so little fat left in the body the meat did no more than scorch on either side. They each took a piece and tried to chew it as it was. Mertz found it repugnant, but kept on chewing. Mawson bravely told himself that, apart from its strange, musty taste, its stringy, sinewy toughness, it was 'quite good.'

It was a forlorn moment, looking into each other's faces, with the taste of a faithful animal on their lips. Mawson eased it with a cracked grin:

'We've a long march ahead, Xavier. We ate
nothing yesterday and what we've had won't
get us far... so, let's have our eight ounces of
hoosh each – and a biscuit to christen our
new eating utensils.'

It was past six in the evening when they finished the
tedium of breaking camp. They were ready to march, to
leave the ragged skeleton of the stripped sledge in the
snow when Mawson was struck by a thought. He said
to Mertz: 'We did not mark our furthest eastern point
yesterday, as I had planned. I forgot it in the shock of
Cherub's death.' Now he cut a further length of wood
from the remnant sledge and nailed the flag he had carried
to its top and set it up in the snow – on the, spot where
he had lanced Ninnis' finger.

He stood to attention, bareheaded and announced: 'I
formally take possession of this new, untrodden land in
the name of the Crown, and I name it – subject to royal
consent – King George V Land.'

They marched away leaving the Union Jack flapping
in a twenty-mile wind. It was the only colour they could
see in the world.

At first, they walked to the north-east, almost back-
tracking, to avoid a heavily crevassed area of ice; after three
miles they turned north-west for two miles, and they set
the line due west, as far as they could judge – for the sky
was overcast and the compass erratic. Mertz and Mawson
pulled with the dogs, Mawson taking up George's position. Uphill, the climb took them to the brow of the

plateau ice, nearly 3,000 feet above sea level; and, as they rose, so the temperature fell to twenty-two degrees below freezing. Mertz started to feel the loss of his waterproof leggings and helmet; the fine snow melted on his head and saturated his woollen balaclava; the denim pants were soaked from plunging into loose snow, and soon his socks were sopping wet beneath his finnesko. Mawson handed him a spare pair of heavy woollen underpants... but while they gave extra warmth, they still did not keep out the wet.

Mawson cheered him: 'Keep going, Xavier. We'll get some good miles behind us and then we'll camp and have a hot hoosh, with fried liver to follow.' At midnight the sun laid a sheen of gold over the southern ice. Then it rose above the dull, heavy clouds and the appalling glare that spread over the land allowed no shape, or design, no feature, to reach their eyes... Mawson walked ahead on the plateau lip, fearing the treacherous surface, virtually feeling his way and often peering under his goggles. He was not to escape the penalty.

A rising, icy wind in the early hours gave Mertz relief; it froze his pants and helmet into solid windproof shields. Mawson talked to help their marching; if they kept to the higher plane and went further south, they would cross the feed-ice to the two great glaciers and there was a chance that the surface would not be so severely crevassed there. It had the disadvantage, he reflected to himself, that the wind and the temperature might be worse... but if they could keep walking, putting the miles away as they were this night – then they had reason for hope.

For ten hours – without halt – they trudged into the west, the sledge meter showing they had covered a marvellous eighteen-and-a-half miles. The reading was then down to eight degrees F... twenty-four below freezing, so they camped, full of anticipation of food and rest.

They fed the remains of George to the dogs, and spent two hours getting their tent erected and the snowblocks cut to hold down its skimpy skirting. They were no sooner settled inside than the glare through which they had trudged took its toll on Mawson. He was in agony with a bad attack of snow-blindness – in both eyes. Mertz tended him, placing the tabloids of cocaine and zinc sulphate under his eyelids. In an hour he could see out of one eye – enough to watch Mertz prepare their supper.

The pemmican hoosh was little thicker than strong tea. Mertz remembered the strong dog, Shackleton, eating their big pack of butter. 'It would make a difference if we had that here now,' he said. Mawson said it would also make a difference if they had Shackleton.

Mertz cooked the liver. He laid it on the tin lid, over a medium flame, and, just like the meat, it gave no fat. It scorched, and did not fry. After a few minutes the outside started to blacken, but when he cut it into two pieces the inside was still pink and raw. 'It is better for you like this – not to cook the goodness from it,' Mawson said. They ate it with their fingers, finding the taste 'undesirable and repellent,' but glad that it was easy to chew and to swallow. Afterwards they boiled the tea-bags, for a third time, drank the weak liquid and prepared for rest. Mawson could not sleep because of the pain in his eyes, and Mertz again dressed them and bandaged his head against all light.

In the small tent they rested through the day, side by side, comforted by the thought of the nourishment the food would give them, unaware they had swallowed the first intake of a poison that would make a nightmare of their days, bring madness and death to one of them, and almost demolish the other.

SEVEN

Diet of Dog

The fight back opened in a strange, spectral scenes An unmoving cloud shut them into a white, hazy capsule through which the night sun glared, hurting their eyes. They trudged each long night – to avoid the drag of soft snow during the day – guessing their westerly path by the south-north alignment of the sastrugi, often buried under coats of snow blown from the plateau. Under the influence of magnetic forces above their heads, and the proximity of the South Magnetic Pole, the compass was useless.

Feeling for the roll of the ice waves, and crossing them at right-angles, was their only guide to the line of march. It made their progress halting.

For three nights they fought these conditions, marching upwards of twelve hours without breaking for refreshment. Mawson aimed at an average of fifteen miles a day, a target that demanded iron will, the desperation of famished men fighting for survival.

They were a sorry sight. Two men, strength flagging, slipping and staggering, falling on the hard ice and cursing the bruising, impatiently wiping the snow drift from their goggles, pulling with five matted, scrawny dogs. Mertz headed this dismal procession, a tattered leader with an old undervest wrapped round his head to retain warmth in his saturated woollen helmet, tied by twenty feet of alpine rope to the front trace of the sledge. He wore skis at first, to sniff out their path between ice hummocks and possible crevasse danger, but finally discarded them as too cumbersome among the sastrugi. With his sticks he prodded suspicious ridges and snow rises, he jumped on snow banks, trod into the troughs, feeling forward, hopefully toward the dream of food, warmth, shelter, somewhere in the west.

Mawson trudged in the dog trace, taking George's place, lugging the sledge through snow and over the ice humps. He walked with his head bent, one eye still bandaged against the attack of snow-blindness — after five applications of cocaine and zinc — concentrating on keeping the sledge in a straight line with the rope attached to Mertz. They moved onwards, at a funereal pace through their weird surroundings, with the swish-swish of the sledge runners, the crunch of boots in crisp snow and the whine of hunger in the panting dogs the only sounds... hour after hour through the nights, long punishing trudging, mile after mile.

Their mouths were parched, their nasal canals dried out; they longed for liquids, but would not stop. To melt the deeply frozen snow would mean lighting the primus, which meant erecting the tent... and that was hours out of their time, and in Mawson's anxious mind hours were miles – and miles were survival. The further they tramped, the more the thought of food intruded, flooded their minds, and – when they rested – dominated their dreams, in vivid, disturbing colour and reality.

They used anticipation as a psychological crutch; they promised themselves a special lunch each seven days – a thin slice of frozen butter, a stick of chocolate, and a boil-up of a tea bag... once a week. Mertz curled up in the corner of their little tent on the third night and wrote: 'How beautiful that will be! After what we now eat, to have butter, chocolate, tea, seems like a dream.' His mind was already questioning the dog flesh. 'In this plight we swallow the dog meat only because we must, our provisions are so scarce. But, while it is better than nothing, it has nothing to recommend it, at all. The nourishment is so small we find it very hard to orient ourselves and do this hard tramping... we lose our balance so often that we fall and stagger like drunkards.'

–

In four nights they covered about sixty miles, but Mawson drew little solace from that fact. The work load was overwhelming for both men and dogs on such a meagre diet. All their strengths were fading fast; and as the dogs died, one by one, the progress would slacken. A daunting prospect, it ate into his optimism: 'An average of fifteen miles a day is critical to our chances, but it is a wretched, trying struggle to cover ground and to keep a straight course, when we don't know whether we are too far north, or to the south. It is a poor outlook for us and a black outlook for the dogs...'

Still, the dogs pressed forward dumbly in their harnesses; their strength decreasing with each mile ahead. Battling, smelly Johnson, was the first to fall, he tottered and his skinny legs buckled under him. He was loaded onto the sledge and dragged through the night to the next morning camp. He did not get to his feet again. They gave him a small piece of George's frozen meat; but he had no power in his jaws to crush it. The fragment lay between his paws and he licked it with a lethargic tongue. It was the end for Johnson. Mertz took the task; he shot the dog through the ear and, sorrowing, skinned the body while it still held some warmth. Old Johnson had a very pronounced odour in life. In death they found it pervaded all his tissues. Mawson tried to disguise it by chopping the muscle fibre into small pieces and stewing it in their thin pemmican. Mertz still found it repellent.

The remaining dogs – quiet Mary; Haldane, looking more than ever like a blue-wolf; Pavlova, and Ginger – showed no such hesitation. They tore the pelt to pieces, crunched the bones, and gulped the offal like starving jackals – even swallowing the teeth whole.

There was little sustenance to be gained from the carcase. By the time they collapsed the huskies were fined down to furry skeletons. They were bereft of fat and protein – yet, with the wicked perversity of fate, their livers were heavily loaded with an excess of subtle but powerful substances, which – while normal to their breed – were biodynamic destroyers in the human metabolism.

Cutting up the carcases, Mawson and Mertz noted how 'relieved' they were when the liver appeared. They relished the freedom from the incessant chew-chew-chewing of the meat; they welcomed the time-saving in cooking, a quick scorch either side and they could swallow chunks – ignoring the fishy, foul flavour and the slimy, clinging texture of the canine liver; they believed it to be nourishing food.

There was nothing to alert them. They had not heard of the lore of Arctic Eskimos that the livers of polar bears or the bearded seal (cousins to the species in Adelie Land, which had been fed to the huskies) were dangerous. They had not read the writings of the great explorer, Nansen, on men who were reported to have died from eating such livers. They had no reason to link the peril with their dogs. However, Mawson knew well of the role of proteins, fats, carbohydrates, sugars and of calories in the diet. He did not know – he could not know – that without the vital balancing effect of the powerful substances that we know today as vitamins none of these foods had much value. The nature of these vitamins, tiny dynamos of life processes, were still on the frontiers of biological science. Indeed, in that very year of 1912 while he and his men sheltered from the great gales at Cape Denison, the word 'vitamine' was coined for the first time in Europe, by Kazimierz Funk.

Not for another eight years would the excessive substances attacking Mawson and Mertz through the dogs' livers be isolated in Britain and named – Vitamin A. Even then, it took a further twenty years before medical science could elucidate the havoc that overdosing of this vitamin created in the human body. When it emerged as a clinical entity it was given the name of Hypervitaminosis A. What it betokened was a biological chaos, a revolution among the normal balanced metabolic agents, an upheaval in human organs, a disruption of other vital vitamins. The symptoms were catalogued: firstly, dizziness, stumbling, nausea; then scaling, splitting of the skin, loss of hair, opening of skin cracks round the mouth, nose, and eyes, later developing into painful, open fissures; the drying out of oral and nasal membranes; then irritability, skeletal and stomach pain – from the chaotic swelling caused in the liver and the spleen – loss of appetite, dysentery, dramatic loss of weight, lassitude, and then morbid sensitivity, irrationality, followed by delirium, dementia, and, finally, convulsions and probable death from brain bleeding.

Still later came awareness that all these afflictions did not arise from the excess of Vitamin A alone. Many symptoms came from the destruction, or disruption, of other vitamins – especially Vitamin C with its role against scurvy – by the overwhelming action in the liver of Vitamin A.

Last of all came the link with the liver of the Greenland husky dog – *canis familiaris*. Not until 1971 was it shown by biochemists in Adelaide that there was peril in eating their liver, an organ that, under evolutionary pressure over many centuries of seal-meat and polar bear diet, had, possibly, built an abnormal capacity for storing this specific vitamin. This work revealed that four ounces of the dog liver contained a toxic dose of Vitamin A for an adult man. Thus, in retrospect, the decision to eat husky livers was to be as catastrophic to the systems of Mertz and Mawson as was the loss of Ninnis with their food down Black Crevasse. Husky livers, it is shown, do not shrink appreciably in size during starvation – they weigh something over two pounds in normal dogs, so that with six dogs to eat, Mawson and Mertz, between them, swallowed some sixty toxic doses of Vitamin A. It was an intake made all the more debilitating and dangerous by the lack of other normal foods that would transport the substances to excretion.

The marks of illness were on their faces within a week of eating the first liver. On 18 December, Mary broke down and was shot, and they gladly ate her liver. Soon they marched through their unreal world in conditions made worse by the liver-induced disorientation of which they were unaware. The parched mouths, the dried-out nasal membranes, the stomach pain that bent them over to one side – all were put down to thirst and hunger. They were never to know that the swelling of their spleens and livers disrupted their balance, that the stripping of skin, the loose shreds of flesh inside their cheeks, the inflamed cracks round their noses and mouths were due to the dog liver and not to wind, cold, and general malnutrition.

There was no way they could know. Their only concern with the dog meat centred on obtaining the maximum benefit, and on how to make the musty flesh more palatable, and easier to chew. Mawson wrote one night: 'We rack our brains on the march trying to think what we can do with the dog flesh to make it more acceptable.'

Slogging on across the ice wastes, feeling the dog harness cutting into his waist, aware of the spreading pain that bent him sideways, he began to think, for the first time, not whether they would get to the hut in time to meet the ship, but whether they would get there at all. The vitamin poisoning was challenging his resolution, eating into optimism; his spirits sagged into gloomy introspection... incessantly keeping the damn rope straight with the sledge, or the other way round, with nothing to see but Mertz's shadowy shape plodding ahead and the dogs stumbling with him... it was a wretched situation... and then, all at once, again, he was flung into sudden crisis and action.

The surface had changed to uneven layers of fresh snow over hard ice. It did not register immediately on his worried mind. Trudging aslant, one eye still bandaged, his good eye lining up the path of the sledge with the line of the rope to Mertz, he was suddenly aware his lead dog, Haldane, had vanished from sight. A yelp of fright and a jerk; the rangy blue-wolf dog was hanging by his tether in a steep-sided crevasse. Instantly, instinctively, Mawson hauled back, stopping the other dogs, and then slowly pulled, leaning backwards, dragging on the trace to bring the frightened, scrabbling dog to the edge of the crumbled snow bridge, hand over hand, until the bullet head was above the lip and the front paws were frantically scratching at the snow to stop his body plunging backwards to doom... and in horror Mawson saw the harness slipping over his shoulders. In one quick move, still holding the trace, he flung himself flat, forward into the snow; his fingers grabbed the hackles, and with a mighty heave and a yank Haldane was hauled to solid ground, squealing with pain and distress. Panting, he lay beside his rescuer in the snow. Mertz came hustling back, concerned. Mawson managed a cracked grin: 'We almost lost a third of our rations, Xavier,' he said.

He got to his feet to fasten the harness on the quivering frame of Haldane: he remembered the busy winter days at the hut, and Cherub Ninnis and Mertz, laughing with each other, stitching, sewing these made-to-measure sledging harnesses that now hung loose on the starved animals.

They moved off slowly and cautiously. After a few hundred yards the sun broke through the clinging cloud and out of the haze emerged the vista of the great river of ice, dipping, rolling like a huge wave formation in front, sliding away to the sea in the north, its roots hidden in the mystery of the inland. They had walked onto the edge of the back of the Ninnis Glacier. And crisscrossing their path, wriggling through the ice ahead were the telltale white lines, ridges of treacherous crevasses barring their way.

For all the danger that lay ahead Mawson was pleased. He commented to Mertz: 'It is comforting to know roughly where we are, to see some feature we can recognise.' Slightly downhill now, the snow-coated ice slipped away beneath their feet to join the main frozen stream. Mawson plugged into this region, giving Mertz a spell with the dogs. When the drift blew knee-high, he felt his way forward by shuffling his feet. They did not get far. Mertz called him back and there was Haldane, wrung out by his frenzy in the crevasse, legs tangled, lying in the snow with his eyes glazed and his tongue lolling. Mawson unhitched him and carried him to the sledge. Poor dumb brutes, he told himself, this is how they all go… pulling, pulling for man – unto death.

There were only two dogs left now, and two men. Haldane's weight slowed them down, and then they were forced to swerve south-west to avoid direct confrontation with an area of rippled crevasses. There was heavy snow between the sastrugi here, and the runners dragged against the surface with leaden heaviness. They made three miles, and then Pavlova fell. Mertz was slumped in his harness, as dejected as the dogs and looking done in; Mawson could not accept this was their day's march, but they all needed cheering.

He called a halt. 'Xavier,' he said, 'tomorrow is the day for our special lunch. But since the air is fairly still we can try to rig a lean-to and light the primus. I think a drink of hot water with a few drops of alcohol will be stimulating.'

Neither of them mentioned it, but they both recalled; tomorrow it would be a week since Ninnis fell to his death.

The special lunch came next day – midsummer day – and then it faded into memory, like the hours that followed and the nights of marching that flowed under their trudging feet.

Haldane had been killed, his body proving the boniest of them all – but there was a large liver. The temperature was down to thirty-one degrees below, and the sledge meter froze solid. The metal had crystallised – when Mawson tried to free it, the spindle snapped in two. This threatened their only means of measuring the distance they marched.

They faced a shattered ice-bed; Pavlova was staggering rather than walking; Mertz was bent over and complaining of stomach pains. They had made only eleven miles, but had to camp where they stood. The snow was near to being ice, it was so cold, and digging the blocks for the skirting of the tent was more backbreaking and laborious than ever. In the middle of the task the shaft of the spade broke in Mertz's hands. They sat on the edge of the sledge, on the lee side, with the tent cover pulled over their heads and the two dogs sheltering at their feet and with frozen fingers repaired the spade handle. It took them two hours.

Holes had to be cut into the frozen snow in which to rest the rough frame, and, recognising their sheer exhaustion, Mawson cut a flat block with sloping sides in one corner of their shelter − lifting this out he excavated underneath and it became a frozen cover for their indoor latrine. He knew their reserve of strength was so depleted they would not make an effort for bodily relief − and that was another health hazard to be avoided. They did not need to crawl from the tent now; they only had to lift the ice-lid and cover it back over. It was in the bottom corner of their narrow little tent and, when not in use, the foot of a sleeping bag would rest over it. Such things were very basic to them now. It matched their approach to harsh reality − like their constant talk of food, for that revived memories of life, colour, warmth, human kind. At times they babbled of places where they had been, of restaurants in which they had eaten sumptuous repasts – of reputed places in Europe...

'There is one such noble dining-room in the hotel across the square from the railway station at Basel; and one evening with my mother and father, my sister and brothers, we held a big family dinner, relatives and friends, and we ate the stag – what you say? – venison... yes, with pommes nouvelles and green petit pois, artichokes, and brandy, and many bottles of Burgundy...'

Mawson had his own memories but wouldn't give them too free a rein. One of his dearest was the last meal at the hut, with his companion gay, smiling broadly in the throng clamouring for his special savoury penguin omelettes... 'I will look forward to a special omelette when we get back, Xavier.' Mertz brightened at the thought. 'Yes, yes! More, more – I have a new idea; I will cook for you a special omelette – I will name it Omelette Mawson... it will be flavoured not with wine, or herbs. It will be flavoured with the whisky from Scotland!' It was a game that gave them support.

Mertz's eyes were suddenly bright with the thought of getting back on board the *Aurora*. 'I cannot imagine how it will feel to sit again at a table – to have a steward to bring my food.' Then he looked seriously at Mawson. 'I have come across the world, dear doctor, with the dogs, with you and the ship. I have seen too much ice and snow; I have glimpsed only a little of the coast of Australia. I want to see Australia and to go to New Zealand – where so many of our comrades come from.'

'Of course,' Mawson assured him. 'I will do all in my power so that you can see my country and travel to New Zealand… I promise you that.'

Mertz did not talk again. He pulled the hood of his bag over his head and thought of the future. Mawson turned his hands to repairing the broken spindle, lighting the primus to heat and reheat the metal, tapping it endlessly with his 'toy' hammer, fusing, burring the broken edges, joining the faces by the insistent rap-rapping. His persistence gave them a serviceable sledge meter, which turned until the final step was made. It was early afternoon before he was finished and he crawled into his bag. The wind was rising and slapping the tent. He could hear Pavlova and Ginger crying for food, and from the restless movement of Mertz' bag he knew his companion, like himself, could not sleep for hunger.

–

It was well past midnight when they fought up a long, sloping snowfield, having taken four hours to break camp. They moved with dreamy lassitude, walking being a conscious effort. They reached the rise of the hill and were both faint and ill from exertion, panting, and their hearts pounding with struggle. Mawson wanted to press on, but they did not make another half mile. Pavlova collapsed, finally. Mawson carried her body to the sledge, strapped it safely aboard and covered it with the tent fabric to retain what warmth was possible… and as he walked back to join Ginger in the traces he could feel her sad eyes following him.

It was a short march. They were among crevasses when the wind shut down visibility with veils of snow drift, forcing them to camp. Gloomy at the few miles covered, sorrowful at the end of Pavlova's usefulness Mawson took the rifle and quickly shot her through the neck. He deliberately set his mind to tackling the pressing matter of the weight they were pulling.

'I tell you,' he declared to Mertz, 'we have to cut down on every ounce we pull. Only one dog now, and she won't last long. All that is not vital will have to go. We have two dog skeletons – bones – in a heavy bag. We'll get rid of them and use a couple of gallons of kerosene at the same time. We'll boil them into a jelly. We'll fry the liver and cut it into squares so that we can eat it cold and so save time in cooking. We have to be rigid, hard about these things – or we are never going to make it back to the hut.'

For hours, he crouched on his haunches over the pot, boiling the bones from Haldane's carcase. Mertz went to sleep. When he woke, Mawson was smashing Pavlova's bones with the spade, outside the tent. He was cold and grim when he crawled in through the tent funnel; he had Pavlova's four paws in his hands. He skinned the fur from them and dropped them into the steaming pot and stirred them with one of their rough wooden spoons. After a while he looked at Mertz, his eyes solemn and bloodshot from the snow-blindness.

He said: 'You know how she got her name, don't you?'

Mertz nodded: 'Yes, I was on the ship.'

Mawson murmured, thoughtfully: 'Of course, I'd forgotten. Anna Pavlova! She gave me that doll, you'll remember – in the cabin on the *Aurora*...'

The warmth of the stove filled the tent. Its heat collected in the peak of the cover and melted the frozen condensation of their breath. Soon it started to fall like raindrops, on to Mawson's head, into the steaming pot. It had been a bright April day, a northern spring day with flying clouds and few scatters of rain; he was aboard the *Aurora* tied up on the Thames, and the huskies had arrived and were quartered on the deck, attracting sightseers along the embankment. He kept a sailor at the gangplank to keep the curious from wandering aboard. Stores were on the deck and he was busy in the cabin with a loading manifest when the gangplank sentry knocked at his door... 'Here's a special lady to see you, sir.'

She seemed to fill his drab little cabin with a radiance – not beautiful, but a striking woman, with poise, who moved with grace and smiled with deep charm when she said her friend Nellie Melba had told her about her very brave young countryman who was to sail a little ship to the bottom of the world.

'I hope you will not mind that I come to see such a ship – and such a man?' Her voice had a lilt that softened her accent. She held out her hand and he held it for a moment, greeting her with a shy deference.

She handed him a wrapped box. 'It is a present for you. Nellie Melba tells me she gave you a subscription to your funds – but you only ask money from Australians; so, this is my gift. I wish it to keep you safe and bring you home again…' He unwrapped the box and found a replica doll, a ballerina on tip-toe, in a blue tutu… 'Take it with you – in place of me,' she smiled.

Mawson was touched, then had his inspiration. 'I wonder if I dare ask… nobody has christened my ship – will you please be godmother to the *Aurora* for her Antarctic adventure?' He found a bottle of wine and they walked across the deck where she poured some wine over the oaken forestem… and asked God to watch over all who sailed in her. They could not waste the remaining wine; they drank it together in his cabin, and then he escorted her to her waiting limousine. On the way she stopped to fondle a young husky, a bundle of brown and ochre fur. Mawson was enchanted, delighted. Anna Pavlova, one of the greatest dancers ever, was godmother to his ship… and, there and then, he named the pup she fondled – Pavlova, the husky, whose paws he was now boiling to jelly, to help keep him and his companion alive.

He broke his line of thought and said to Mertz: 'This bone stew will set quite thick, Xavier. Surprising the food value you can get from things…' His eyes settled on Mertz's wet fur boots, the reindeer-skin finnesko. 'I'd not be surprised to get some food value out of boiling them down,' he said to his astonished companion. 'After all, the dogs have eaten them.' Mawson was again forcing his mind to the harsh facts of economy, extracting every possible ounce of energy from their resources, and always coming back to how they could divest themselves of weight… 'We'll go through all that we have; as soon as we can, we'll try to lighten the sledge. Only us and one dog...' He went on stirring the dog paws long after Mertz fell asleep.

-

They woke at eight in the morning on Christmas Eve, stayed their hunger with a piece of cold, scorched dog liver and a boil-up of an old tea-bag. Mawson had a renewal of urgent energy. He hurried Mertz into the march, telling him they were many miles behind schedule, that rations would have to be reduced much further if they could not keep advancing.

'It's a bad state of affairs,' he growled. 'Snow has fallen and made the going heavy, and we're not getting stronger.' Pointing to the sledge load: 'Some of these things will have to go.'

Again, his mind was dwelling on how to move faster across the ice. What was really essential to their survival? What could safely, usefully, be dumped? He wrestled with the problem as they slid and slipped through the snowy blanket lying on the back of the Ninnis Glacier. There were problems in the energy-sapping drag of the runners through the soft surfaces, there was time lost in making and breaking camp... how to lighten the load? It was to lead him into an error of judgment that came near to having fatal consequences.

They made only three miles by noon. Mawson decided to camp and settle the question burdening his mind. The load of the sledge was laid upon the snow. The box camera, all the heavy glass plates would be thrown away, he decided. Why should they carry a rifle – and bullets?

165

Other instruments, the hypsometer – which he had used religiously each day to plot the altitude of land – the thermometers, old socks and unwanted almanacks and log books – other than their diaries – could all go. The heavy pieces of sledge runners used for their tent props would also be discarded. They could be replaced with the telescopic legs of the theodolite – and that instrument in its box was heavy, but he could not navigate without its help. And all the bones would be boiled down to jelly and dumped – that very day.

More hours over the primus, stirring, waiting for the marrow to stew down. When it was done, the snow was sheeting horizontally in the wind; he rolled into his sleeping-bag and at once an ocean of tiredness washed like waves over his mind, drowning out his practical thoughts and bringing the sharp longing for nourishing food into the fabric of his sleep...

–

His Christmas dream was a colourful, rich and tormenting flight of fantasy. He could see himself, free from troubles, walking a snowy street, with windows full of candles on decorated trees, rooms gay with bunting, and sounds of music... and then he was in this wonderful, glorious, confectionery store – a wide, sweeping emporium filled with the greatest cakes that ever were made... large creations several feet in diameter, enormous cakes piled with sugar icing of ravishing colours, mouth-watering immensities in sponge, with whipped cream, and fruit flans, and huge custards... and he had money and could make his own choice – a wonderful, thrilling moment to look at the cartwheel of cake that would be his own… Then he was out in the street again, black despair flooding him. He had paid his cash and left the store – without his cake. The windows were now in darkness; there was a notice on the door which glared at him: 'CLOSED!'

–

It was 11 o'clock when he woke; a thin light filtered green into the small tent. He lit the primus and started to heat some dog stew.

While the dog brew was steaming he ferreted in a corner of his bag and found the half biscuit he'd hoarded for this day. He woke Mertz, handed him his food and then broke the half-biscuit into two pieces. 'A relic of better days, Xavier,' he grinned. It was now Christmas morning. They supped their hot liquid stew, guarded every morsel and crumb of the precious biscuit. Mawson squatted in the corner, Mertz on his elbow, back against the snowy tent cover; two men, alone in the world, with the wind flapping the cover above their heads, half-blinded from snow glare and their eyes running a watery fluid, their faces cracking into painful smiles – they wished each other happier Christmases in the future, and again they talked of food.

Mertz recalled it was two weeks since Ninnis died. The memory seemed to lay a sense of doom over him and he fell silent, wondering at his premonition. All through their slow, fumbling preparations for the trail he remained wordless – depression flowing through him.

They trudged through their Christmas scene of snow showers for seven hours, and finally camped at 9.30 am. They had their celebration meal – a small knob of butter dropped into the dog stew. The sun glimmered to greet them and Mawson at once went out to try to fix their position. The legs of the theodolite were now tent props. He put the instrument on the cooker box, but found his hands trembling badly – unable to hold them steady, he scribbled the figures and clambered back into the tent.

His recording was meticulous – as it had been all through the trip: cloud formations and direction, wind strength, temperature, altitude, position, latitude and longitude. Only the details of coastal alignment and character remained lodged in his memory – everything else was written down, day by day.

'We are some eighteen miles further south on the glacier than when we were here a month ago. I'd say we are roughly eighty-five miles from The Crater – where Ginger ran away? I saw a whale bird while I was outside, a prion; surprising distance from the coast, some hundred miles I should think… and if we had wings like that bird and could fly direct, the hut would be 186 miles from here…'

So they had come about half-way considering the diversions forced by the terrain. There was another great glacier to surmount; and they would not again make long marches as on those first few nights. There were questions in his mind – how far could they go on the food they had… but it was now Christmas, their Christmas together. He said to Mertz: 'Let us thank God, Xavier, that we are still alive. Let's promise to remember each other on this day for the rest of our lives – and to celebrate this day together when we are in better circumstances.'

Mertz brightened: 'Yes, yes. When we get to the hut I will make your whisky Omelette Mawson! We will have fine dinners... and never again will I eat the dog.'

Before they set off again Mawson pre-empted their next rations. He opened a new food bag and made a thin cocoa with dried milk. They ate some more liver, then Mawson held his wet woollen helmet over the primus flame to dry out some of the moisture... he heard Mertz ejaculate: '*Warten Sie*... a moment, still, please!' He plucked at Mawson's left ear. His fingers lifted away a complete skin cast. The entire covering of the ear was drooping from Mertz's thumb and forefinger.

Alarmed, Mawson fumbled his right ear – and again the spread of skin came away. Mertz took off his own wet helmet, and, inside the wool, strips of skin and small tufts of beard were attached; there were raw patches at his temples, and his hair line had receded. Mawson noted his companion's once-luxuriant black moustache was patchy, ragged, that the cracks round the mouth, nose and eyes – which he had put down to wind and cold – were opening into red, raw fissures, deep, like razor cuts, and runny.

Was this condition only on their heads? They released the bottoms of their pants and over their boots, on to the snowy floor fell a small stream of skin strips and loose hair.

'I think our bodies are starting to rot,' he told Mertz. 'It is lack of nutrient. We are walking, working hard, wearing out our tissues – and there is little nourishment our bodies can find to do the repair work.'

Mertz's eyes were dark with returning presentiment. 'I have felt all the time,' he said gloomily, 'that the diet of dog does not agree with me.'

EIGHT

Cross in the Snow

The nights following Christmas were filled with the drudging toil of climbing out of the Ninnis Glacier. The struggle to reach the 3,000-feet high plateau of ice on the western side carved deeply into their reserves of strength, and on the third night they sorrowed at the loss of the last husky.

Ginger fell into a tangled heap with fifteen miles still to go before they cleared the glacial fringe. Full of mingled pity and admiration, Mawson laid her on the sledge as she quivered with the last dregs of energy. He stroked her head and covered her with a bag, and, as her glazed eyes watched him walk back to his sledging harness, Mawson remembered how this remarkable animal had run back to their previous camp through The Crater on the outward journey. She had now outstayed, outlasted by days, the biggest and strongest dogs of the team.

She rode the sledge as a passenger for three miles, and when they camped it was obvious she would never walk again. Now she had to serve a further purpose – as bone jelly, meat and liver. The rifle had been dumped to save weight; the spade was the only weapon and Mertz could not face the task. Mawson was compelled to deal the death blow, breaking Ginger's neck. 'A pitiful thing to finish off such a fine animal in such a way.'

Mertz helped with the butchering remembering as he did the long weeks at sea with Ninnis tending all these charges. Once the tent was up he felt exhausted, dizzy and sick; silent, he went directly to his sleeping-bag, there to record the day's events in a brief note... ending unknowingly with a paraphrase of the words of Captain Scott written that same year on the Ross Shelf... 'I am sorry, but I don't think I can write any more.'

-

In the early morning the wind blew sheets of snow against the tent. Mawson ached for food, but first answered the constant nag on his mind – the weight of the sledge; the bones would first be boiled down to marrow jelly, and then 'the skeleton could be safely jettisoned.' After simmering out the skimpy nutriment from Ginger's depleted marrow, one source of food was left that he could not ignore – Ginger's head.

It made a scene that burned into his memory. Two haggard, tattered men, crouched in the narrow tent, watching the skinned dog's head cooking. They had no way of cleaving the skull in two portions; so they discussed, and agreed, how it should be shared; the sledgers' method of 'shut eye' would be used. Normally, a hoosh cook would serve out roughly equal helpings and the other men would turn their heads, close their eyes and the cook would ask – 'Whose?' – touching one finger on one helping.

When Ginger's head had boiled for ninety minutes, they lifted it, with the two impromptu wooden spoons, onto the lid of the cooker. Mawson ran a knife across the top of the skull for the demarcation line; Mertz closed his eyes and Mawson touched one side of the head and asked: 'Whose?' They took turns in gnawing their different sides, biting away the jaw muscles, lips, swallowing the eyelids and gulping down the eyeballs. With the wooden spoons they scooped out the contents of the skull – and then split the tongue, the thyroids, the brain into two servings and again employed the 'shut eye' method; only their fierce hunger made them grateful for this macabre feast, only the pangs of starvation made it possible for Mawson to write before sleeping: 'Had a good breakfast from Ginger's skull – ate brains, thyroid and all.'

The devouring of Ginger left a new loneliness in both men. They missed her presence and were more aware of the weight of the sledge on their aching bodies. As well, her end came at the time when Mertz showed the first clear signs of decline. Mawson noted, almost suddenly, that his cheerful and non-complaining sledging mate was morose, downcast and depressed:

> 'Xavier is off colour. We did a good fifteen
> miles, halting at 9 am, but he turned in at
> once, all his things very wet on account of his
> lost waterproofs. This fine drift penetrates all
> our clothing and doesn't give us a chance to
> dry things. Our gear is deplorable.'

With so far to go to safety and so little food there was dark foreboding in Mertz's dwindling health. Yet, Mawson would not allow his mind to preview disaster, but fixed his attention on how to travel faster, go further, lighten the sledge? In the centre of this problem was the 'wretched, makeshift tent'. The crude structure was a tormenting waste of time. He wrote of his frustration: 'Eight to ten hours straight-out pulling is a great strain on two men, which could be avoided if only we could pitch the tent halfway through and halt for a boil-up... but to halt in our poor condition in the open would be to freeze. There is no alternative to this slow, ceaseless, seemingly endless pulling – hour after hour.'

Very soon Mawson would be grateful for such marches, for, never again would they trudge such distances together. Dragged down by his dreadful illness, Mertz sought the fatal alternative to the endless slogging; to lie in his sleeping-bag and wait for ideal weather. On the evening of the first day of the year, he was already committed to this policy of inaction. With his cold-bitten hands he wrote in his diary:

> 'January 1. New Year; 5 pm. Only made five miles before having to camp and am now back in my sleeping-bag. The light is terrible, the sky clouded. We could not get far in such weather. It is best to wait for sunshine. I cannot eat of the dogs any longer. Yesterday the flesh made me feel very sick.'

They were the last words Mertz wrote. He remained in his bag for three days, sinking deeper into despair, gloom, and periodic irritation. Mawson tried to persuade him to make an attempt to walk. He placated and tolerated his mounting revulsion against dog flesh by making over their entire stock of powdered milk, only to find Mertz turn away from food entirely. He even rejected an offering of their 'supreme luxury', a biscuit and a knob of their last piece of butter.

It was part of an inexplicable pattern of confusing moods, which marked the wasted hours while the blizzard drift whirled above their tent. Trapped on the exposed ice-hill, unable to move, Mawson's anxiety deepened. He tried to build his companion's fading strength, and out of kindness fed him a half-share of his own piece of Ginger's liver, arguing in his compassion that it would be easily chewed and digested... Mertz slipped rapidly into deeper lassitude, into disregard of his plight. Mawson's talk of meals they would eat, of omelettes at the hut, of fine London restaurants was useless. So, he filled the hours of dark gloom with dressing Mertz's splitting lips and the raw patches on his legs and groin with lanolin from the medical kit, persuading him to swallow hot drinks of cocoa and tea and, between times, stitching and repairing their clothes and drying socks over the primus flame. Nothing halted the progress of the illness and the slide into weakness of body and spirit. Desperate, Mawson induced him to swallow some dog stew, but, suddenly aware of the taste, Mertz flew into a feverish rage and flung the tin pannikin across the tent, into the snow. 'It is of the dogs,' he yelled. 'They make me ill because I eat their flesh!'

On the third afternoon of the year, Mawson enthused to the sick man: 'Look, Xavier. Your sun has come! You said you believe in omens, remember? Here is the sun, calling you to march to the hut.'

He aided him out of his bag, and dressed him in his sledging gear. Mertz sat on the sledge as Mawson collapsed the tent and packed for the trail. Hopefully, with better visibility, they could make some distance... but, again, they made only a faltering, winding trek, slipping and falling, and after four miles Mertz stopped, hands hanging at his side, panting heavily, head drooped. His fingers were again badly bitten by frost. They had to camp. He sank into ominous silence huddled in the wet bag; Mawson, perturbed, heated some cocoa and ate a piece of Ginger's liver. In his diary he entered:

'Xavier is in a very bad state; everything now
depends on Providence.'

In the following two days Mawson felt his own strength slipping away, fast. He was nauseated, had pain in his side, and all movement made him dizzy. Acutely anxious at the passing time and their wasting food, he pressed Mertz constantly to resume marching. Mertz would not accept the argument that to rest in the sleeping-bag and wait for good weather was to wait for death on the ice. Mawson wrote in his diary: 'He practically refused my request that he should make an attempt to walk – even for a mile or two. He told me it would be suicide and that he would die if he tried to march in bad weather...' They had a short argument over the question of moving for the sake of keeping their blood flowing, then Mertz sank into gloom. He refused again to march when on 5 January the sun broke through, but Mawson won his promise that he would walk next day, if the conditions were good.

January 6 brought cloud, a twenty-five-mile wind and drift, but passable marching weather. At 7 am, Mawson climbed from the tent to start preparations for the trail. It was a three-hour struggle; he had to dress Mertz, feed him cocoa and biscuit, and help him don the sledging harness. The depredation of cold, hunger, exposure carved their faces, Mawson knew, but, once out of the softening green light of the tent, in the stark cruel light of the plateau, Mertz's appearance was shocking. His beard was ragged tufts with patches of raw skin beneath, his moustache was a travesty and his eyes deep-sunk pockets of grey beneath the protective goggles.

–

The march was a shambling, fearful trial of flesh and spirit. Weak and trembling they felt their way through poor light in a slow meandering of which Mawson wrote: 'The surface was slippery and we both had frequent falls bruising our emaciated bodies. Quite dizzy from the long stay in our bags, I felt weak from the lack of food. But – to my surprise Xavier soon caved in. He went only two miles with many long halts, and then he refused to go any further...'

Mawson was aghast at the implication. He threw an arm round the bent shoulders to help his comrade forward but Mertz slumped into the snow: 'My mind goes forward, but my legs stay here,' he moaned. Mawson hauled him to his feet. 'We must go on! Our lives are at stake. We must walk while we can.'

Dr Xavier Mertz was beyond the reach of logic – beyond words or thoughts of a frozen death. He stood dumbly. Mawson sought anxiously to avoid facing the awful fact that his companion was near the end. A week had gone by with few miles marched. A daily average of the ground they had covered in that time wouldn't save them. What could he do – his own body weakened by starvation and privation? Certainly, he could not just camp and capitulate. No! He would not just lie down in the snow and wait out more days of agony as life slipped away.

'Get on the sledge, Xavier. The way is downhill and we can go a little further, and perhaps you'll feel like walking again.' Mertz resisted. There was some affront for him riding the sledge, but Mawson insisted, pushing him onto the load, making him lie down, covering him with the sleeping-bags.

The canvas harness cut into his shoulders. He leaned forward to haul the sledge and the leather belt dug into the painful area of his right side, so he canted to the left. He could feel the draining drag of the runners through the snow and the sudden judder as they struck the sastrugi and rode over the ice-backs. He staggered as his feet slid on the wind-polished surfaces and he heard Mertz moan with pain on the jolting sledge. The fearful thought crept into his mind – a sudden slip, a broken leg, or ankle, and they would face the end... better to go down on all fours, on hands and knees. I can pull with the best of dogs, he told himself; I will do as well as Ginger... I can go on, and on! The pounding of the steely ice jarred his knees, and he felt the skin split; snow crammed his gloves and crept into his clothing. He kept on crawling, lugging the weight across the snow, his whole concentration focused on making distance.

He covered two and a half miles this way. Then Mertz was calling out, in pain. He was very cold, with white spots of frost on his paper-brown cheeks. His condition was piteous; he uttered no word of complaint, but when Mawson helped him from the sledge he seemed unwilling to move his legs.

Mawson managed to erect the tent and get him inside within an hour; then he heated a thick cocoa and dog-stew, which he pressed on Mertz with the name of 'beef tea.' The name deceived Mertz's mind and he drank the hot liquid; yet, his stomach soon rebelled and he vomited into the snow. Anguish was a ball of lead inside Mawson. In his journal Mawson expressed his anxiety:

> 'Things are in a most serious state for both of
> us. If he cannot go on and make eight or ten
> miles a day in the next day or two, then we
> are both doomed.
>   'I could perhaps pull through – with the
> provisions at hand – but I cannot leave him.
> His heart seems to have gone. It is very hard

179

on me; to be within a hundred miles of the
hut and in such a position is awful.'

Mertz seemed to fight out of his dark depression. He said he was sorry they had to stop. 'How far did we go?' And then, brightening, he said he would ride the sledge again in the morning and perhaps they could put up a sail with the tent… There was no more. He lapsed again into silence. Mawson turned in at 8 pm, to face a 'long and wearisome night'. His last words for that day were: 'If only I could get on! But, I must stop with Xavier. He does not appear to be improving. Both our chances are going now.'

–

Mawson woke from troubled dreaming of food to find the prospect of travel a shattered hope. Mertz was in a dire state. His trousers were fouled from an attack of dysentery, his eyes were wild and rolling, and he talked and babbled incoherently; he sounded demented. Mawson set to work cleaning Mertz's soiled clothing and was shocked to see his legs, his groin, stripped clean of skin, red, raw, rippled with painful folds. It was a long and tedious task before Mertz was restored to his sleeping-bag. Mawson felt the cold badly and crawled back into his bag for an hour to try to recapture some warmth. At 10 am, that morning of 7 January, he again rose when Mertz started shouting. His companion sat up, obsessed with a thought that Mawson wanted him to get out and ride on the sledge; his eyes were glaring.

Mawson was moved with deep pity. The affable, philosophical companion and friend had vanished, given way to this poor incontinent wretch, a shattered man who feared to move. Yet, even in his deep compassion, Mawson could not imagine the terror Mertz might feel in riding the sledge as a passenger, that it could be an awful omen – a ride to slaughter, as it had been for each of the beloved dogs. Mertz raged: 'Am I a man – or a dog? You think I have no courage because I cannot walk – but I show you, I show...' He lifted his left hand. The little finger – yellowed from frostbite – was thrust into his mouth and Mawson watched in stupefied horror as Mertz crunched his teeth into the middle joint, savagely severing the skin, cartilage and sinew, tearing away with grimaces and groans of pain... then, in disdain, spitting the severed digit into the floor of the tent, the trace of thin coagulated blood turning the snow pink.

It opened a day of madness, of raving, and constant fits. Mawson dressed the stump of the finger and wrapped it in the, bandages Mertz had used for his snow-blindness... He now knew that Mertz would never walk again. Outside the tent the sun was showing through broken clouds; the conditions were good for marching:

> 'Obviously we can't go... this is terrible! I don't mind for myself, but it is for Paquita – for all others that are connected with the expedition – that I feel so deeply, and so sinfully.
>
> 'I pray to God to help us.'

The babbling fits, the incoherent German and English, were followed by a quiet spell when Mawson again fed him cocoa. In the evening the raging became violence. Flinging his arms about, thrashing across the small space, Mertz broke one of their precious tent props and would have caused further damage but for Mawson sitting on his chest, holding his arms, struggling to quieten his dementia. Dysentery attacked him again, and he fell into unconsciousness. Mawson again cleaned him, and his clothing. Once more, Mertz broke into raving; holding the side of his head he lay back in his bag, calling over, and over: 'Ohren, ohren! Ohrenweh!' ('Ears, ears! Ear-ache!')

The terrible day was done. At midnight Xavier Mertz lay in a coma. Mawson softly toggled the skin of the sleeping-bag under his chin and wrapped the soft fur round his raw, skinned face. Drained physically and emotionally he crawled into his own damp bag seeking relief in sleep.

His rest was troubled by disquiet and at 2 am, he woke. In dazed perplexity he searched for a reason for waking; there was no movement, no sound other than the ceaseless rattling tent. He reached out to touch his companion – and Xavier Mertz was stiff, cold, quite lifeless under his hand.

The awful truth was a blanket of cold fear, invisible, but falling over his entire world, filling the tent, flooding his mind with the terrible, haunting fact. He was alone. All that was human in this accursed place, all that had been alive – friends and dogs – were dead, and gone. Loneliness was in the vast wasted land outside, in the soughing wind, in the corners of his mind, in his anguish and in the fear for his own safety... he was himself, sick, famished and so weak he might collapse at any moment; and he lay stretched out on this floor of snow with the heart-rending truth pinning down his body and his mind. Mertz was dead.

What had gone wrong? What would he do? What chance had he of living? Very little, he decided. This spot was some hundred miles direct to the hut; ahead ranged the heaving wind-swept plateau ice, the great, broadly-fractured bed of the glacier, many miles of wicked winding crevasses; and then the long grinding, back-breaking climbs up the steep slopes and ice-ramparts to the escarpment near The Crater... to be in sight of Aurora Peak, to leave some record there where they might come seeking his missing party. Yet, he was so emaciated that the bitten, snow-clad peak seemed a million miles away...

Mertz was dead by his side. Why then was he alive? In that question were hours of heart-searching, restless harking back over the weeks gone past... they had eaten, slogged the trail together, suffered the same cold and lack of food... and there were deep, hidden roots for the sense of sin that swept his being: fine, noble young men, and he had led them to their deaths. He could phrase a justification in his mind: 'The accident to Ninnis, and its consequences are like fortunes of war – things always liable to happen, a risk that is part of the game and inseparable from pioneering in an ice-covered land.' But, why had

Mertz died? Death was due to cold, constant weather exposure, and the effects of starvation culminating in fever and convulsions... that *had* to be the diagnosis. He could find no other. And there was relief in the thought that Xavier was now beyond the reach of pain and suffering. He could compose a eulogy for a comrade he had come to love like a brother... 'Surely he has broken free from the fetters of this icy Antarctic plateau and has gone where a high mind and sterling quality meet true reward... Myself, I seem to stand on the wide lonely shores of the world with only a short step into my own unknown future...'

The outlook was gloomy. The gnawing pain in the stomach seemed to have developed into a permanent companion so that he could no longer stand erect... how could he pitch the tent singlehanded in the strong winds?

In the early morning he heated some cocoa, and, covering Mertz's ravaged face with the sleeping-bag cowl, he wrote an entry in his diary:

> 'For many days – since 1 January – Xavier's condition has prevented us going on; now I am afraid it has cooked my chances altogether – even of making a single journey north to the coast. Lying in the damp bag for a week on extremely low rations has reduced my condition seriously. However, I shall make an attempt. I shall do my utmost – to the last – for the sake of Paquita and all the members and supporters of the expedition, to at least get some word through on how matters stand with me.'

To get far enough toward Aurora Peak, for his body and their diaries to be found, that was the aim. Over the slopes, crossing the glacier, up the terraces and ramparts on to the escarpment and past The Crater…

> 'Today I shall spend remodelling my gear for
> one-man travel. I shall cut down the sledge to
> carry half the load and I shall doctor my own
> worn body and broken skin as best I can.'

It was time for action, a way to face the empty, lonely wasteland and to meet its trials.

He sat cross-legged in the tent for hours that morning. Mertz's waterproof coat was cut and trimmed and stitched into an old food bag to form a sail. Outside, defying the flying drift and wind, with his little hand-saw, he cut the sledge down to half its length and used parts of the discarded frame to mount a small mast and spar to carry his waterproof – thus windproof – sail.

More hours of crouching over the primus stove, cooking the last meat from the dogs, scorching the remaining half of Ginger's liver. Reduce the need to cook down to the minimum, he argued, and you cut down the weight of fuel you need to carry on the sledge. In fact, he found he could leave two one-gallon cans behind.

In the afternoon lull the plateau wind fell away to a chill breeze, and he then faced his final duty to Dr Xavier Mertz. On pages torn from his notebook he wrote his statement attesting Mertz's uncomplaining bravery, how the death of Ninnis had led to this second fatality and how he himself was continuing westward. The note was placed in the sleeping-bag, along with ten dozen exposed photographic plates, now accounted too heavy to carry in the final, bitter struggle toward Aurora Peak. He dragged the body from the tent.

The wind was rising again as he laboured to cut the snow-blocks for Xavier's burial cairn. The sky grew dark and threatening; he left the corpse in the stained sleeping-bag – a shroud of reindeer skin – and soon drift and falling snowflakes covered it with purest white.

Building the cairn to shoulder level made his heart pound like a trip-hammer; he rested often. When it was done he took out his prayer-book and, for the second time on the journey, stood bareheaded in the snow and read the burial service. They had stood together some four weeks before, and he had then named the Ninnis Glacier. The daunting glacier he had yet to cross would now carry the name on the world's maps – if his body and their diaries were ever found – of the Mertz Glacier.

He was dropping with fatigue, needing food and hot drink, but then it came to his mind that a fast-moving party of fit men might cross that gulf of country between him and the hut; there was need to mark the death cairn. The two halved runners, cut from the sledge, lay by the tent. He thrust them into the top of the cairn to form a simple cross in the snow.

NINE

One Pair of Feet

The tent leaned drunkenly. Tiny in the white, horizonless immensity, stained and drab green, it sagged over the support broken in Xavier Mertz's final raving. Standing at the side of the burial cairn, bowed with sorrow, Mawson looked at his frail shelter and, in its crookedness, saw a symbol of Mertz's suffering – and of his own vulnerability.

He crawled into the tent, utterly weary, aching from the toil of the burial, and his stomach creased with the pain he put down to hunger. He felt a deep desire for hot, richly fatty food, and for a long warm sleep that would refresh his body and wash away fears of the perils ahead and the hopelessness of his plight. The wind was now rising and the snowfall was thicker. The tent cover was flapping and the broken ends of the strut were shaking; he forced himself to repair the fractured prop. If the ends pierced his cover, if the wind could rip through a tear, his shelter would soon be gone; and without it he would be finished. He fitted the little hammer-head to the Bonzer hand-knife, foraged for nails and lamp wick lashing in the repair bag, and set about splicing and binding the broken support.

His fingers fumbled, his hands trembled working above his head as he squatted in the snow; it was troublesome work, but it was nostrum to his anxiety. When he had finished, he slumped where he sat; his whole body yearned to be in the reindeer bag, not to be moving, and he knew he could not yield. There were other tasks to claim his mind: light the primus and heat a pan of pemmican hoosh with dog-bone jelly, sew up the tears in his clothing, count up his food supplies – and write the diary entry. The last helped him to compose his mind:

'I read the burial service over poor Xavier...' and at once he was projected into his own predicament and the looming peril:

> 'As there is now little chance of my reaching
> human aid alive, I greatly regret inability to
> set down the coast line as surveyed for the
> 300 miles we travelled; and the notes on
> glaciers and ice formations, most of which
> is, of course, committed to my head.'

Time, his own collapsing condition, the reduced food reserve all urged him to make an early start on his long journey west... should he march in brief spells, or just go on until he fell in his tracks? When and how soon should he pack and go?

The continent answered that question. A bustling river of deeply cold air burst down from the great icy canopy; booming gusts of wind above fifty-mile strength soon surged over him and hurled a torrent of snow drift and flying crystals at his tent. The tent frame quivered and shook so menacingly that he squatted in the snow with his hands above his head to hold the struts into their ice-holes.

Endlessly, it seemed, the gale stabbed and roared round his shelter and he feared its raking blast might reach under the skirting and lift his cover from the snow-grip and fling it to some unreachable distance. His hands were shot through with pain, his arms ached with leaden throbbing; not until late evening did it come to him to lash the struts with lampwick and hold them down with the weight of his body. It was some relief, but now his inert frame was filled with lassitude, and deep malaise flooded his mind with the crushing burdens of his problems; how could he hope, with this weakened body, to break camp, to march and erect the wretched tent in such wind? He was under pressure to yield, to surrender, to eat, rest easy, until the final sleep closed over his eyes with peace. Somewhere deep in the well of his being the primitive, powerful urge to survive, to exist at all costs, exerted pressure on his subconscious and raked his tiring memory for words of comfort, for motivation, and there came lines written by another man in a different world – to gird his resolve, to aid his fight – words from Robert Service:

'Buck up! Do your damndest and fight;

It's the plugging away that will win you the day.'

It rose in his consciousness like a call; it lifted him from despondency and was a challenge to his character... to attain some further westward point, to get beyond Aurora Peak where his body and their records might be found, to tell the tale of his comrades and himself, for Paquita, for the search parties he was sure would be sent out to find him while the ship waited to sail west to relieve Frank Wild and his group. His love, his men, his ship and expedition; he owed them all a duty.

Buck up! Do your damndest, and fight. That was a call to his character – and – by God! – he had pontificated enough about character when he'd been promoting support for this expedition: the meetings of scientific groups, the public appeals; the huge audience in the brightly lit Melbourne Town Hall, resplendent for the occasion, the Governor-General, Lord Denman, glinting with decorations, Prime Minister Andrew Fisher, the Federal Opposition Leader, Alfred Deakin... pomp and power of the land under the one roof and he could hear again his own voice:

> 'I have done my best to choose men of
> character. The important thing to look for
> in members of an expedition like this – is
> character. It is impossible to tell how men
> are going to act until circumstances arise ...
> In that land of desolation, in that land of
> great loneliness, there are the conditions that
> measure a man at his true worth.'

Character – when your own death looms over the tent, when all is against you, when all that seems sensible is to lie down and wait to die. It's the plugging away that will win the day – but outside the world was a chaos of drift flying in the wind, denying passage to the fittest, strongest, and the most resolute of men. The tent rocked and juddered under the assault. 'I am in the hands of Providence!' he consoled himself.

All the while the fear of losing his cover occupied his mind, and when the wind fell away later in the evening, he crawled outside to cut more snow-blocks to heap on the skirting. The snow was frozen, hard to pierce, and as he strained to lift one block a sudden squall took him off balance – and the spade handle broke in his hands.

Nearly four weeks had gone since the time Mertz had sat with these same tools over this same job – after the long march back from Black Crevasse. That same day they had first eaten the flesh from the dogs, and he had assessed their food resources and plotted a ration to take them near to succour. Now he repeated the tedious tacking and splicing and lashing of the handle – the same assessment of his pitiful stock of food. Nine weeks he had been in the field and for the last four weeks had been near the starvation level of eight ounces of dried food a day, plus the dog flesh. The life-saving food bag that had been on his forward sledge that day of the Ninnis tragedy had held a sledging diet of thirty-four ounces a day per man – a total weight of forty-two pounds of dried food, plus the remaining dog meat, the bag of raisins and almonds and the box of chocolate sticks which had been in his personal bag – the only such bag not to be lost in Black Crevasse. Mawson could not weigh what he had now; he could only assess that he might make this scanty provision last for a given time – and, so he asked how long he could last? How far could he go on this diet of about eight ounces a day? Say twenty-five miles to cross the glacier, twenty miles to Aurora Peak and then the great curving ice dome beyond... and the allowance gave him twenty days of subsistence for the fight across the ice. Say, again – two biscuits a day; two little chocolate sticks and a few raisins as refreshment and thirst quenchers on the trail; thin pemmican and cocoa twice a day, supplemented by stews, the dog fragments, a small piece of liver, the sinews, and a set of paws.

What if the primus stove should break down? He would be dependent on provisions that did not need cooking; so, he divided his food in two sections: that which had to be cooked – pemmican, cocoa, dogmeat – for the first ten days; and the biscuits, the bone jelly, half the chocolate and the little bag of raisins and almonds for the second half.

Now his life was sectioned into twenty days – up to the end of the month. There was no future after that. And if he could march an average of five miles a day – taking all weather into account – he could come near to where he might find help. But where exactly was he now?

The sun glimmered through cloud in the evening sky; he set the theodolite on the cooker box. It was hard to get the instrument level, yet he felt the result to be near enough to put him roughly a hundred air miles from winter quarters, from the hut... but, he had no wings; and he would walk many additional miles, wandering through the glacier and over the difficult land beyond. Still, it was a target. It answered his iron resolve to get as far as he could – on what he had. It was grist for his deeper wish to survive.

He returned to his repair of the spade handle and when the sky cleared and the sun lit his tent it roused a desperate urge to pack and walk – if only for a mile or two while the weather break lasted. Somehow, he could not move. The spade had to be mended, and he argued that his raw scrotum, his stripped legs would be better for another night's healing rest... and, as well, he feared a return of the wind. Buck up! Do your damndest and fight! Yes, but, with this poor body, how could he manage the toil of camping in a fifty-mile wind, of securing the sledge, carting the gear inside, lighting the stove... with these bitten fingers?

They poked through the ends of his tattered woollen mitts – blackened at the tips from frost-bruising, cracked round all the nails and festering on some. And what of the rest of this once-powerful body? He untied his lampwick braces and belt, and lowered his thick underpants – and a shower of skin fragments and hair fell into the snow at his feet. Strips of skin had completely vanished from his legs; his knee-caps were without cover, just roughly rippled flesh, his private parts were red raw, scarified from the friction of work and walking. Around his waist, on his shoulders, the harness had laid a skinned pattern; and here and there he found eruptions breaking out like small boils, with festering heads.

'My whole body is rotting from the want of proper nourishment,' he lamented. 'There is nothing to repair or replace my worn tissues.'

The weather promised fair for the morning so he turned in early. In the period before sleep his mind ranged across the facts of his predicament, and suddenly he no longer felt he was alone, but that a presence was with him. A few yards away Xavier Mertz rested beneath the burial cairn, but Mawson felt his spirit still in the tent, by his side.

–

He woke to the gift of a peaceful morning: sunny, almost calm, with only wisps of cirrus cloud flying to the northern sky. Ice and snow-fields stretched around in all directions into white-hazed horizons. In the still air it was very cold. He did not want to risk being chilled before breaking camp so he went back into his bag again for an hour until the sunlight gained more strength. Every fibre in his being longed for him to be moving again. He counselled himself: 'I must go slowly, steadily. I will not make a long day of today, but even so if the surface is good perhaps I might win ten miles.' In three days he had walked no further than the cairn. In the week before that, in the trauma of Mertz's sickness and dying, they had done no more than a dozen miles… in a dozen days! The need to march weighed on him, but his caution warned that his body would need time to be moving freely again, to unloosen his joints, to regain the rhythm of marching – and of pulling the sledge, alone.

He rose at 8.30 am, and it took him more than two hours to heat a thin hoosh, collapse his tent, and lash and pack his possessions aboard the foreshortened sledge. The old food-bag with Mertz's converted windproof stitched into it, was tied to the mast and the spar; the cooker-box was loaded and he was ready to go.

First he stood for a while by the cairn to pay last respects to the remains of 'a brave gentleman and a fine companion', to Herr Dr Xavier Mertz. He attached this formality to the brief act at the side of the cold grave and murmured a short prayer for the dead man's soul – then added a plea for his own safety.

He turned away, donned the sledge harness and, bending to his left to ease the pain in his side, he pulled his load slowly onto a downhill slope. The slight breeze was from behind and he felt its movement ease the pressure on his body. After trudging about one hundred steps he stopped and looked back, a deep sadness on him at leaving the forsaken, desolate spot where Mertz was buried. The death cairn was now rounded by the night winds, the cracks between blocks were filled with blown snow and the whole had the shape of an elongated igloo. The cross of the two halved sledge runners stood stark against the frozen wilderness.

His trail led away to the north-west; soon, the cairn was lost to his view, for ever. Very deliberately Mawson restricted his gait, shambling more than walking, leaning into the harness rather than pulling. Overhead the sky draped a peacock blue that fell into a misty meeting with the distant ice. The sun was clearly visible and its warmth on his face stirred an idea in his mind. After covering about a mile, he halted and filled his cooker tin with snow, then strapped it onto the sledge where the sunlight would fall on its outside and melt the contents to drinking water – to ease his thirst as he marched. At the same halt he rested the theodolite on the cooker box and took a reading to find his course heading was forty-four degrees west of north. He had two sticks of chocolate in his pouch, and he ate one of these as he resumed his pulling, moving into his north-west trail.

Every step was painful; the raw areas between his groins, the skin-stripped scrotum, induced him to walk with his legs apart. It shortened his step but he found that on the snow-covered ice it made his tread more secure, and in this way he walked another downhill mile. The hard backs of sastrugi started to show from beneath the snowy cover; he slipped and staggered and in the jarring of his legs he felt a new disturbing pain in his feet spreading discomfort through his ankles into his legs. Persisting with the target of ten miles that day he went down on all fours and crawled across the ice waves. The new pressure and the pounding on his knees wiped away the new pain for a while and masked his mind against the awkward, lumpy, squelching feeling in his feet, as though he was treading in treacle; there was also a deep-rooted ache in his ankles. Soon a new sense of apprehension filled his mind; the sun was bright and there was little wind and he stopped again, to find the cause of this fresh affliction. He sat on the edge of the sledge and took off his finnesko and the two pairs of socks.

The sight of his feet was a hammerblow to the heart. The lumpy, awkward feeling came from underneath – where both his soles had separated into casts of dead skin.

The thick pads of the feet had come away leaving abraded, raw tissue. His soles and heels were stripped; an abundant watery fluid filled his socks, and it was that which had caused the squelching feeling. A wave of despair rode over him. He sat aghast, staring at the ruined feet he had trusted to carry him to Aurora Peak. He was to write later: 'All that could be done was to heavily smear the red inflamed exposed flesh with lanolin – and luckily I had a good supply – and then replace the separated soles and bind them into position with bandages. They were the softest things I had available to put next to the raw tissue.' He took all the socks in his bag – six pairs in all – and pulled them over the bandages, then forced his soft finnesko over the top.

The shock told heavily on him. He could not at once start marching again. He felt the sun's light falling on his face and found it refreshing; suddenly he was seized with the idea that solar energy would benefit his body. He spread his sail in the snow and took down his trousers, pulled up his vest and, defying the cold air, lay down to expose his body to the light streaming from the sky. Later that night he noted: 'I bathed in the glorious kiss of the sun and an almost instant feeling of well-being went tingling through my body. It was exhilarating, a sensation that flooded my senses.' The sunlight's energy seemed to glow under his skin, to stimulate restorative processes that eased the deep anxiety roused by the state of his feet, and which made him feel stronger and more confident.

He lay in the sun's light as long as he could bear the cold. Then he covered his body and renewed walking, avoiding hard ice where he could and treading into soft snow. Each step was a controlled movement to avoid too much pressure on his feet. Sometimes he walked on the outside edge of the finnesko, sometimes reaching on to his toes. In between he dropped to his hands and knees to rest his feet. Through the afternoon he continued, pausing to sip a drink from the melting snow in his cooker, nibbling his second stick of chocolate, edging his way north-west.

The pain wore deeply into his nerves. In the early evening, though the air was clear and the sun still shone, he gave in and camped. By the reading of the sledge meter he had trodden six and a quarter miles that day of 11 January. His strength was exhausted; he scribbled into his diary: 'I am nerve-worn from the pain in my feet. Had I gone any further I should not have found strength to erect the tent.'

It took him all of ninety minutes to set up his shelter and stow his gear ready for cooking his pemmican and the last set of dog paws – he no longer knew from which animal they came. 'With long stewing,' he noted, 'they came down to digestible softness.' The hours of this evening were given over mainly to attention to his body. Unwrapping his sore feet, he found them in a 'much more deplorable condition'. The trouble had spread round his ankles. 'The whole skin has burst into blisters – almost the whole of both feet.' As well, he noticed his toes were blackened from frost bite and, like his fingers, were also festering. He spent hours with the dressings on his body, and bandaging his loose skin soles back into place.

As he ate his meal he became aware that he could no longer taste food and had lost his sense of smell. 'The membranes of my nasal canals have gone wrong and my saliva glands are refusing duty – for want of proper nourishment.' But, despite the lack of taste and smell, his hunger was razor-edged and he looked at his food stock and wished intently that it was twice the size... 'Then I could set to and have a fine meal, whether I would taste it or not.' The longing now was to feel the bulk and the warmth in his stomach and his only satisfaction came in the vivid dreams of food that pursued him in sleep.

–

The sun shone brightly through the tent fabric. He crawled outside to enjoy the light... 'I am sure the sunshine will have set much right in me; I felt the good of the sun as I have never done before. Sunshine is the elixir of life for those who have been without it for weeks, never taking off their clothing in the snows and winds of the Antarctic plateau. I am determined I will be a sun-worshipper – for as long as I can live!' The thought helped him back to the reality of his plight.

He stood by his tent in the dazzling white scene, the land dipping downhill away from him – into a valley of ice. All at once he saw he was on the edge of the frozen chaos of the main stream of the great glacier he had named for Mertz, with its daunting upheavals and depressions, and there, some thirty miles further on – rising in the cold hazy air – he saw the dark rocky summit of Aurora Peak. And beyond that, he knew, the great ice plateau rose and rolled away into distance – to Aladdin's Cave, and the hut. He lifted his face toward the sun and a silent prayer formed in him:

> *'Oh! If Providence will give me twenty days of weather like this and will heal my feet – surely I can reach succour.'*

TEN

Corpse in a Crevasse

For more than thirty hours a snow-laden, angry wind had lashed his camp. Now, suddenly, sharply, it had passed. It was ominous silence, the quiet of the grave, and Douglas Mawson sat bolt upright in his tent, his ears straining, his mind hungry for sound. No wind, no movement, not even the rustle of whispering snowflakes broke the frozen calm.

He put his head through the tent entrance and saw cold, featureless desolation and in this setting, in this instant, his solitude became utterly consuming. The feeling of the presence of Xavier Mertz had gone; loneliness was a chilling fact, almost a pain in itself. He thought: 'Oh God! I could be all alone in this whole world, a poor creature, a solitary man of pre-Cambrian times… or a lost soul on the surface of Mars!'

His sense of isolation increased and weighed more heavily on his spirit when he found he did not know the real time. He had again forgotten to wind his watch.

He guessed the time to be about 7 am, but until he could sight the sun that was uncertain. His world outside the tent was filled with diffuse and treacherous light; the land was exactly the same colour as the sky and conditions were too dangerous for him to march. Impatient as he was to be mobile, he had to bide his time and to use work for his hands as a panacea against worry, against the anxiety of the burdensome hours of waiting and knowing the passing of time reduced his food supply – and his chances. So, he re-bandaged his sore feet, dressed the inflamed areas of his body with soothing lanolin, applied iodine to the eruptions and to his festering fingers; he cooked a half-ration of pemmican and then, hoping to build strength for the journeying ahead, added some of his dwindling stock of dog jelly with half a biscuit; then he boiled up an old tea bag. He re-lashed the tent prop and tightened the binding on the spade handle; he dressed himself for the trail, packing a layer of dried grass into his finnesko to ease the pressure on his feet... all done steadily, to while away the hours as he waited for the sun, hoping the enforced rest would help to heal his tattered body.

On this fuzzy, bleak Monday 13 January, the sun broke through a little after midday, and he balanced his theodolite on the cooker box, took his readings and found his watch to be two hours forty minutes fast. By the time this was done it was 1 pm. Overhead the clouds were fragmenting and the growing strength of the sun was dispersing the haze, promising a clear, calm after-noon. Hurriedly he broke camp; soon after 2 pm, he tied the twenty-feet alpine rope, knotted every three feet, to the back of his harness, adjusted his goggles over his snow helmet, and, leaning forward to take the drag of the runners on his shoulders and midriff, he moved down into the tortured valley of the glacier, facing many days of glaring light, snow falls, and showers of frozen crystals – and, underfoot, the hidden perils of snow-topped crevasses and sudden, sheer precipices in the, hard, blue ice.

Almost at once the downhill run of the glacial ice broke into sharp-edged corrugations, ridges that butted through the protection of his fur boots, dry grass, socks and bandaging. He tried all his tricks of walking, shuffling flat-footed, on the sides of his soles, bending out the ankles – and the iron-hard ice still took its toll; he could feel the abraded skin bursting, and the fluid filling his socks. The pain and the fear of severe damage forced him to seek patches of virgin snow… and he knew he invited danger. 'I am just blundering forward,' he told himself. Yet, he had no other course but direct across the Mertz glacial valley. Fit and well-provisioned he would have detoured far to the southern highlands to escape the trials in front of him; he knew full well the awful slog through the rising terraces, mile after mile of cruel uplifting ice, of the heart-pounding ramparts at the glacier's western banks… that afterwards there would come the vast, curving, gale-swept plateau, times when he would fight the wind to erect his tent, days when he would be blizzard-bound.

In mid-afternoon the sky cleared as though some magical hand swept the cloud away to the north. Risking the blight of snow-blindness he lifted his goggles from his straining eyes… and an awesome landscape rolled before him. At once his heart was uplifted. In the far distance beyond the western banks of the Mertz Glacier the black-topped rocky mass of Aurora Peak was sharp against a sky of pale blue; it called him on and stirred him to fresh effort. It was a signpost to winter quarters, he had told his now-dead companions. Now it was symbolic to his struggling body, a citadel to be achieved.

But was there an easier road than this hazardous direct course? In the south the highlands rose in massive banks and he could see the sun glinting gold on the frozen falls that fed the glacier. Too far, too exhausting for his frail body! Could he go down the glacier, follow the rolling banks of ice downhill, until he reached the sea? There he might find food, seal meat, or penguin, or eggs, and regain some strength. But would there be any likelihood the search parties would come seeking him along the coast? He could see no real option to his chosen path through this deadly region of crevassing.

He battled on ahead. Now and again he stopped to sip water from the snow melting in his cooker; he nibbled on a chocolate stick. Down, and still downward the course ran. With each shambling step forward he felt the immensity of the river of ice cracking and grinding its way to the sea, of the titanic power needed to push the huge floating tongue, some 2,000 feet thick, across the bedrock until sixty miles or so offshore it floated. He was puny against this vast setting, but will power pushed him onward, step after step; sometimes he fell to hands and knees when the ice-backs were too sharp for his feet.

In the low light of evening, shadows sharpened the contours and etched out the ridges of the snow-lids above a network of crevasses. He paused and stared unbelieving at his watch. It was 8 pm; he had walked for six hours covering a distance of five and a half miles. In the deceptive light Aurora Peak seemed no nearer to his eye than when he first saw it loom through the distance. Rather than face the criss-cross peril of the crevasse network in this light, he decided to camp where he stood, among the snow-filled mouths of the icy pits.

There was very little wind, so that, with frequent rests, he had his tent pitched and was inside cooking his supper by 10 pm – pemmican and his last piece of dog liver, with biscuit and cocoa. He gulped the food and only when the last trace was licked from his pannikin did he tend to his wounds. When he pulled off his boots and socks he was again downcast. All the rest he had given his feet was undone, all the treatment had been nullified. The mess in his socks, the blood-soaked bandages appalled him; the raw areas had abraded and bled and copious fluid flowed from his insteps. He cleaned his feet and rebound them for the night. In his diary he wrote of his concern:

'My feet are worse than ever – very painful.
Things look bad – but, I shall persevere.'

Every bone and sinew in his being ached for the peace of sleep in the reindeer bag, but, again, the glacier denied him.

Soon after 11 pm – as though resenting the weight of his wasting body on its mighty back – the river of ice opened a resounding cannonade. Booming, echoing volleys of sharp explosions went reverberating down the length of the glacier bed and Mawson could feel the solid ice under his snowy floor rebound and quake with the fierce eruptions. Disturbed by the outburst, overpowering to him in his loneliness, he looked out from his shelter to peer into the night scene. In the south the sun rimmed the highlands toward which the Mertz Glacier serpented… and from there the outbursts seemed to roll down toward the sea, passing under him. He believed there was large scale cracking and a rending of the body of the glacier. He reasoned that enormous pressure from the vast plateau and the result of the sun's warmth on the back of the frozen river was the cause. He wrote his observations: 'In the chill cold of the night it seemed that great volumes of compressed air were released with every explosion… due to the splitting of the ice.'

Into the night the frozen bombardment rocked the ice under his sleeping-bag, and that night he did not hunger for sound. In the small hours of the morning the wind took over with forty-five-mile gusts bringing flying drift in whirling clouds that killed his hope for an early start into the depths of the glacier's valley.

The wind held him down until early next afternoon. He spent the time in fixing his position by dead reckoning and calculated he was some eighty-two air miles from Aladdin's Cave, south of the headland they had discovered on their way out, and south-east of the vast depression he had named The Crater. He cut his rations that morning because he could not march; he had half a helping of pemmican and a half of a biscuit.

When the sun broke through and the wind dropped, he resumed his winding course through the shattered ice.

Snow walls, banks of drift, small patches of white cover were all potential traps; he carried a tent prop to probe his forward path and it was slow, grinding progress. When he came to hard, exposed ice he donned a pair of crampons to aid his footing and made some headway – but that did not last long. In the still air the sun's light caused little trickles of water, which ran into the cracks, and turned the snow into the consistency of heavy mud. The weight of the sledge, its drag through the half-frozen snow made him giddy with exertion. Exhausted after six hours, he had covered a bare five miles, and could go no further that day.

Weakness, debility, the pain of starvation all attacked his determination again, and when he saw the state of his feet he considered a more perilous course: 'If my feet do not improve I must turn down the bed of the glacier – to try and reach the sea.' Wisely, he slept on the idea.

–

The fracturing explosions in the glacier again broke his sleep. This night he questioned whether the refreezing theory was correct. There could be far more complex causes. 'The sounds are so loud I think some other cause might be involved. The noise resembles explosions of heavy guns, but they come at random, and usually start high up in the glacier and end down toward the sea. There is no real visible evidence of how they are produced.'

The frozen bombardment caused him to sleep late. When he woke it was an overcast morning, at 9.30 am, and at once came the stabbing realisation: 'We should all be back at the hut by now.' This was the deadline date – 15 January – end of the second week in the month, and the *Aurora* by now should have pulled up the anchor and be sailing out of Commonwealth Bay, the main base party aboard, and on her way to pick up Frank Wild and his group, to return to Hobart. He had made the date firm in his declaration at the hut. How well he could recall that evening in the lamplight, the atmosphere full of expectation of high adventure, and serious exploration: 'At all costs we have to be ready to go home by then!' What had happened to them all? Had their plans crashed – like his own – into disaster and death? He hoped above all that none of them had known the cruel agony of hopelessness. And when none of his party appeared... what would they do? How long would Davis keep the *Aurora* there? How long dare he wait to sail to Wild and avoid being trapped in the pack-ice? In his illness, and his hunger, obligation crowded down on his conscience: he was leader, founder, promoter of this expedition. He had brought all these fine men to this terrible land... please, God! Let them all be safely back. May none of them have been hungry, and short of food as he had been – and poor Mertz and Cherub! There was a further rack in his helplessness... what could he do for them now? Buck up – do your damndest and fight! Yes; it would be folly to seek an easier path down the winding bed of the glacier... who would come looking along that part of the coast? He must struggle on – to leave his record, his testimony, near Aurora Peak!

Concern forced him into rash, urgent action. He broke camp and tried to walk into the snowy morass; the soft mess underfoot and the wind beat down his resolve to make another five miles that day. He was stopped in his tracks after two hours of hauling, amid signs of snow-coated crevasses. Another wasted day hung over him; two miles of slogging and he was again compelled to pitch his tent, to lie in his bag and wait for evening to bring the cold that would make the going easier. Also, he had to steel his mind against temptation to eat a warming pannikin of hoosh... not to march was not to eat and in thirty hours he had come a stone's throw of two miles!

Evening brought further disappointment. When the glacier opened its nightly bombardment he went outside for a few minutes; he came back to make a diary entry: '10 pm. Snow still soft. Much snow falling. Impossible to see crevasses or steer a correct course.' At 2 am, it was the same, and in the end his restlessness for distance brought him from his bag at 5 am, when he cooked a weak pemmican and packed his sledge. The wind now blew in cat's paws from the south-east and the sky was heavily overcast, the air full of snow flurries. The light was bad for travelling, and in normal circumstances he would not have budged. Now he donned his harness and trod into the new snow, desperate to cover ground, to try to get free from the menacing clutch of this glacier.

The conditions were abominable. The glare was such that no detail of the ice or snow could be distinguished. The soft snow built into soggy plates over the sledge runners. It formed a clogging ball over the measuring

meter and the turning wheel at the rear of his load...
again, and again, he stopped to free his gear from the
cloying snow; frequently he clambered over drift heaps,
or felt his way along ice ridges that were the lips of
crevasses, and all the while he prodded and probed with
his tent prop, feeling a blind way through a hazardous,
shattered area. 'I believe that Providence walks with me,'
he exclaimed, after several escapes from danger.

Once he felt the shock of falling feet first through
the snow; luckily it was frozen below the surface and,
because his outstretched arms and the tent prop kept him
above ground, he was able to clamber out. He would not
rest while the compulsion to keep going west burned in
him... until 3 pm.

He had fought his way forward for eight taxing hours,
switching directions when the crevasses were too wide or
too dangerous, virtually edging his way across the frozen
valley. The light south-east breeze had all day filled his
sail with no more than gentle pressure, and he came up
a slope of hard ice, onto a rolling brow in the glacial bed
and then there came the first sudden surge of a battering
wind from the south-west. It took his sledge from behind
him, swung it across the ice, round him and to his front,
the tow rope catching his knees and bringing him down.
Both sledge and himself were being blown down the slope
when he dug his crampons into the ice and checked his
own descent. Then, horror-struck, he could see his load,
all his hope of life, all his possessions, teetering on the
open maw of a great hole in the hard blue ice. It had steep
precipitous sides, like a frigid quarry, dropping down out

of his sight. Instantly he hauled on the tow rope and his sledge was braked on the very lip of the ice with the wind still yanking its weight against his straining arms... for a full minute he could do no more than hold the weight. It was eternity to him. Then, slowly, he hauled back, and, inch by inch, the sledge came against the wind, up the slope. He backtracked to a flat snowy area and, with many rests, put up his tent.

–

Exhausted, and faint, he lay on the snowy floor for a while before lighting his stove and preparing his food. This evening the need of nourishment was a sharp and vivid pain, and his only palliative was to crouch over the cooker and see the steam rise from the weak pemmican mixed with a little dog jelly; so acute was his hunger, his crying desire for food, he drooled as he ate the tasteless, odourless mixture.

He was shaken by toil and by his narrow escape from disaster, yet, he had to recapture in his diary the moment when his sledge 'was a yard from the edge of a great yawning crevasse'. His hand shook as he scribbled with his pencil: 'I don't know what lies on ahead at all... I do hope the sky will clear and that frost will come. It takes quite a while dressing my feet each day now.'

He ended his entry there for 16 January: the meticulous notes on meteorology, wind direction and strength, temperature and altitude, the cloud formations – all omitted now in this bitter fight to win ground against the glacier. He had forgotten his heartfelt prayer for fair days;

now he spent restless nights and met and accepted dismal overcast weather as the normal thing... he was coming closer to Adelie Land; and he no longer prayed for good conditions.

On the morning of 17 January, ignoring the falling snow and the virtual white-out, he tramped on a course twenty degrees west of north, grimly set on covering at least another five miles. A plain of ice and snow, on the spine of the Mertz Glacier, rose in front of him; he could feel the ascent, but not see it. The pressure underfoot was on his toes, as, bent almost double, he carted his burden into the morning that would bring him into the greatest peril of his life.

–

He toiled a long, rising slope, heavily covered with snow. The sun was hidden, but its light and warmth filtered the low cloud; he took off his waterproof jacket for easier movement, and, along with his gloves, tied it on the back of the sledge. He strained his eyes to find the safest path in the horrible, deceptive glare. Several times he stopped short of open-mouthed crevasses; twice he actually scraped past gaping cracks he had not seen. He then came on smooth snow and the sledge was running well, when – without any sign – he went through to his thighs. He clambered out with some effort and resumed his climb up the slope. Peering out from under his goggles he made out the line of the crevasse on the edge of which he had just fallen through. It went to the south beyond vision; he turned to the north, and, fifty yards on, all trace had vanished into a field of flat, clear snow which offered him a path back to his westering course.

In the next instant he felt himself falling, his stomach a plummeting lead weight. Then the rope yanked viciously, cutting the harness into his body, bringing a sea of bright-coloured pain; he was suspended over a black, bottomless chasm. Now he could feel the sledge, pulled by his weight, sliding across the snow, toward the edge of this icy pit – nearer and nearer. In seconds the bulk of the sledge would rush over the broken snow-bridge and then he would fall into the abyss... The thought flashed to his mind: 'So – this is the end!'

The movement stopped. Against some unseen ridge, or roll of snowdrift, the sledge halted, and now he swung fourteen feet down, between sheer walls of steely blue ice, six feet apart.

Slowly, he spun in the crevasse, drooping with despair, at the end of the rope. Above, the lowering sky was a narrow band of light; below him were unseen black depths. Cautiously lifting his arms, he could just touch the crevasse walls; smooth and cold, they offered no finger-hold. Overhead the light showed the line of the rope cutting deep into the broken snow bridge, and he was fearful that sudden movement could again start the sledge sliding toward the edge. He held his position; the sledge did not move when he swung his legs in a wide arc... grat-itude filled his heart: 'God has given me another chance...' A small, slim chance. The rope and the sledge held him fast, and the rope was a way out. Yet, how could he haul his weight directly upward on fourteen feet of rope, with his bare hands, his clothing full of snow, his body weak from starvation? Despairing, he turned his mind to the sledge propped in the snow above; how much did it weigh? Would it hold his weight if he tried to climb? He pictured the possessions on the abbreviated sledge and instantly he saw the bag of food stacked on the mid-platform and in the fear that clouded his brain he knew that he must make every effort to reach the bag.

The thought of wasted food galvanised him to action, and he was reaching a long skinny arm above his head, closing his bare fingers around the first knot in the rope. Shutting his mind against pain and stress, he lunged upward with his other hand and pulled his chin level. Again, the reach – and he was six feet nearer the ledge; once more, and then again, holding the rope between his knees, feeling for the knots with his feet now... and he was level with the broken snow-bridge. The treacherous, compacted snow was crumbling; several times he tried to crawl to safety... and he was half-way to solid ice when the whole ledge fragmented under him. Again he crashed to the full length of the rope.

Once more the sledge held its grip in the snow. Once more he dangled, limp, drained, suspended in the chill half-light. His hands were bleeding, all the skin of his palms had gone, his fingertips were black and his body was freezing fast from the snow clogging his clothing, the deep cold of the ice walls shutting him in. He asked – why just hang here waiting for a frozen death, why not end it all, quickly; be done with the pain, the suffering, the struggle? Later, he would write: 'It was a moment of

rare temptation, to quit small things for great, to pass from petty exploration of this world to vaster worlds beyond...'
At the back of his belt was the razor-sharp sheath-knife. A good slash, a moment or two of breathless rush, and, then, final peace... and no one would ever know how it ended, what had happened to him. He could see the sorrowing face of his beloved Paquita, the faces of his comrades... and he pictured again the food waiting on the surface... and Robert Service – Buck up! Do your damndest and fight. Try again!

His strength was draining fast, he was growing deadly cold... soon it would be all over and done with. But, Providence still had him at the end of the rope that was a way back to the surface. By what he later called a 'supreme effort,' he scaled the rope, knot after knot, and, with a wild flailing kick, thrust himself into the snow, above the solid ice. He fell into a faint and lay unconscious, his face toward the sky, his hands bleeding into the snow.

For the rest of his life he could not recall how he made that final climb from the crevasse, nor did he truly know how long he lay unconscious – he believed it to be well over an hour. When he came back to awareness it was in answer to the eye of the sun staring down at him from a clearing sky.

He camped near the edge of the cold pit that had almost claimed him. He could not march. His battered hands fumbled and trembled as he lashed the rough frame together, and, as he struggled to draw the cover over his shelter, he thanked Providence, again and again, for restoring him to life, to his mission.

He filled his pannikin with hot steaming hoosh and cooked some fragments of dog meat and ate it all with a biscuit, still shaking from the experience, living it over in his mind. Sitting in his bag and with the stub of pencil held in his painful fingers he wrote his story of that day, and ended: 'It is impossible... the light gives no chance and I sincerely hope that something will happen to change the state of the weather, else how can I manage to keep up my average. I trust in Providence, however, who has already helped me so many times.'

From now on he felt that spirit always with him, a presence that moved across the snow and ice. He believed then, and ever after, that without divine inspiration the story of his eastern party would have closed with Cherub Ninnis dead and mangled in Black Crevasse, Xavier in his icy tomb, and himself a frozen corpse dangling in a crevasse in the heart of the cruel Mertz Glacier.

ELEVEN

A Coloured Bag

The cold of near-death stayed with him. The hours of the long night were unending misery. Shivering, though he wore all his clothes and his wind-proofs, lying curled up in his sleeping bag, trembling from the exertion, he lived the horror of the crevasse repeatedly. He could not keep still; he could not sleep. Shaking, trembling, he lay awake through the night, hearing the wild wind bluster down the valley of the glacier from the heights of the immense inland dome of ice, from the plateau south of Adelie Land.

He was prey to desperate thoughts. As the hard snow rapped his tent he pictured the glacier outside, with whirling white drift washing over the twisted network of awful crevasses that writhed across his course, and he felt his end in that terrible, desolate gloom was inevitable. There was no pathway; the virgin snow was treacherous. He had taken the wrong course in the fight back from Black Crevasse: 'Had I known this awful overcast and the drift were to be expected, I would have made for the sea and taken our chances there...' It was futile, now, to struggle on. He was trapped in the dangerous ice of the glacier heartland and he was denied proper light to see the perils. 'Light! Light! Give me light!', he prayed. 'If only I had light, from the sky, from the sun, I could fight out of this hole... perhaps I could even make it to the hut.'

The winds from the Adelie Land plateau held out little hope for this dream. Had the sky been clear he could not have seen the sun through the torrent of drift blown over his head. In his weakness his imagination chased the prospect of an easy end to the struggle rather than to go on against cold, distance, wind, in the fractured ice, for the sake of leaving a record to tell his story... a cache that in all probability would never be found. 'Enjoy life for a few days more, sleeping and eating my fill until the food ran out – or to plug on in hunger with the prospect of plunging again at any moment into eternity... without that great pleasure and luxury of having eaten the food – that was the problem that occupied my mind.'

Stark facts seemed to urge the easy course; among the crevasses in this foul light death waited on every step. His physical condition was distressing; his strength was ebbing fast. The damage to his hands was horrendous, the skin ripped from the palms, the fingernails black and loosened, festerings open and suppurating. Dressings of lanolin and bandages only hid the wounds on his body; there was no food, no nourishment to promote healing. He turned for consolation to his diary. 'Things look bad, very bad. I don't think I can endure such strain again. Hard work is beyond me now.'

His diary replaced his companions; each day since Xavier had perished he filled the pages as though talking to himself, weighing his chances, recounting his experiences. In tightly jammed lines, to make the most of space, he was writing longer, taking more words to tell the tale of his trial, and to fill some of the long, lonely hours.

Hope dwindled near to dying in that long night. Always he was forced to the wish that tomorrow would bring good conditions... tomorrow, and tomorrow. The wind racked his shelter and reminded him – he was in the kingdom of blizzards!

Wind was as bad an enemy as overcast cloud and poor light. He asked himself again – how could he hope to travel, to fight the wind, to lash the frame and erect the tent, cut the snow-blocks and unload the sledge, with these broken hands? He sat in his reindeer bag, unwound the bandages and peered in the gloom at his painful, swollen fingers. He lay one palm across the other – and in the crossing of the two sets of fingers there was born the idea that was to save his life. Sprung from the human genius for invention – it brought hope. He could fashion a rope ladder! With his spare lengths of alpine cord and Xavier Mertz's old towing line he could knot and fasten a safety device, so that when he plunged into the next crevasse he could scale the ladder of rope back to the surface. He knew how to tie the loops of rope, how to fasten them to make steps, how to cut a strip of wood from the sledge platform and – making it strong enough to take his weight – use it to separate the two sides of his ladder.

He told himself: 'One end will be tied to the bow of the sledge and I can carry the front end over my left shoulder but have it loosely fastened to my sledging harness.' He was certain it would work – always provided the sledge propped in the snow and held his weight. He went to work at once knotting the ropes, fastening the loops with bindings of lamp wick, stitching the ends securely.

By 10 am next morning, he was ready, and when a glimmer of light pierced the leaden glare he broke camp, packed his gear and bravely stepped forward into the fractured ice. He did not go far. The tumbled upheaval of pressure-ice cascaded across his path; pinnacles, sharp spears of frozen water, jagged edges ripped from the frigid flow by the slow, remorseless power of the vast moving glacier, all barred his way. He pulled toward a slope going steeply uphill, but the cracked and shattered surface caused a wild zig-zag path that soon left him bewildered as to distance or direction. This was perilous terrain for a team of fit men; he could not face it in his starved condition, trudging on raw feet with his mind calling the hazard, urging him – Go back! Go back! But to turn down the hill again was defeat, a conflict with the driving willpower that thrust him forward. Unwilling to face the anguish of retreat he sat for a while on the edge of the sledge, knowing full well that not to keep moving in this wind and cold was itself a danger.

He made only a few steps down the slope when the ground broke under him and once more the fearful sensation of falling flooded his being; he dropped into a snow-masked crevasse, and again the iron-hard rope smashed up under his heart and knocked the breath from his body. For minutes he hung on the rope recovering from the shock; then he took hold of his rope ladder and – full of apprehension – tugged to see if it would take his weight. He could sense the nose of the sledge butting into the soft, snowy surface. For the first time, the thick drift snow, blown down the glacier by the plateau winds, was a blessing. The sledge held firm. Taking his time, with slow steps, he made his way from the chasm. From now on some of the horror of falling to his death vanished; the only question remaining was whether the sledge would hold, and he felt that where snow hid the peril, it would also provide a brake on a sliding sledge.

He shook the loose snow from his clothing and moved into the downhill slope, thinking to skirt the chaotic ice ahead. A few more steps and once more he plunged between icy walls amid showers of fragmenting snow from the broken lid. Again he climbed out... his safety device, he told himself, was a ladder to life! Nevertheless, he was exhausted and he leaned against the sledge to catch his breath, to still his pounding heart. For a brief moment the sun's light found a way between the overhang, and, there, looking slightly east of north, he saw The Crater. There was a flash of hurt; in that vast frozen arena, 800 feet deep, he had halted Mertz from ski-ing after the runaway Ginger; on that crater floor Ginger Bitch had delivered her fourteen offspring, later to die herself, with Ninnis, in Black Crevasse. Heartening was the thought: 'The Crater appears to be as much as five degrees east of north. If this is correct it may mean that I am a mile or two further on than I thought.' And – soaring somewhere to west-north-west – lost in the leaden glare would be Aurora Peak, perhaps a good day's march away! He probed a flat area of snow with a tent prop, found solid base and pitched his tent. He could go no further that day.

The glacier still resented his intrusion. It cracked and splintered; he heard muffled growls and sharp reports, grinding, and then felt trembling, even jerky motions in the floor under his tent. The river of ice he had named the Mertz Glacier was then in audible motion, flowing down to the sea. During the whole night it quivered, like a living thing, frequently waking him from sleep with sudden outbursts.

At 4 am, he saw the disc of the sun through the tent fabric, but, when he looked out, a wind of thirty-mile strength blew drift across the glacier. He remained idle until 7 am, and then could wait no more. He was packed and was on the trail again by 8.30 am; the light and the drift were discouraging arid, beneath him, the glacier still boomed and shook. He was still in the maze of crevasses, some of them with mouths more than eighty feet wide, others closed with hard compacted snow. With faith in his rope ladder, he thrust bravely through these. After about two miles of trudging, back and forth, skirting the worst places, he felt faint; his pulse was rapid, his breath coming in gasps. He stopped to consider what to do, to try to clear his head of the fog of illness and debility. 'Everything seems hopeless,' he despaired. 'It seems impossible for me – all alone – to cross this terrible terrain, expecting any moment to go plunging down!' His illness was again assaulting his resolve, but from somewhere he dredged new heart once more. 'I will go on!' he pledged himself. 'I will stick to this present course as much as possible. I will press on and rely on the goodness of Providence!' The presence he called Providence gave him boldness to plunge on. He was not reckless, but he took chances. He tested the snow-filled mouths of crevasses by stamping his feet, confident now that if he did go through, and the sledge held, he could climb out.

—

Through the hours the bad light played tricks. It gave him no outlines, it cast shadows where none should be; it masked the fatal ridges where snow joined the steeply plunging ice breaks. Still, he pushed forward, through the hours of the morning, making twenty yards, weaving back ten, advancing on a different tack – always toward a general westering direction. About an hour after midday the pale sun unfurled the scene ahead of him.

Two or three miles ahead the western bank of the ice stream fell into a marginal depression, a shallow valley tangled with open crevasses... but, beyond that, starting with a line of ice mounds which he came to call 'pimples' – all had their own crevasses – and rising steeply, was the snow-coated ice-cap, the frozen canopy that covered the land that led on to Commonwealth Bay.

At last, at long last, he could see the end of the frozen hell he had named the Mertz Glacier, and again his heart filled with hope.

He camped early that afternoon. He had trudged three miles and about 200 yards – and each of the added yards was precious. As ever, he wished deeply for better light, for seeing conditions in the morning so that an early start would take him to the beginning of the slogging ordeal of scaling the plateau ice. He held no illusions on the toil it would demand to climb the slopes to an altitude of more than 3,000 feet.

He prepared for this assault as best he could. He scraped and polished the sledge runners, with a small tin of wax carried in the repair bag; he mended tears in his coat, and then, all at once, the sun was bright outside his tent and he took off all his clothes and again spread the sail in the snow and lay there naked so the light could reach his legs, his feet, his skinning, raw patches.

He stuck the cold as long as he could and went back into the tent to dress all his damaged parts with lanolin, to bandage his loose soles back on his feet, to wrap up his abraded hands. When he came to write his diary entry for that night – 19 January – he recalled it was a year to the day since he had farewelled the *Aurora* in Commonwealth Bay, on its way west with Frank Wild and his men. It was, at first, a warming notion that the ship would now be back at anchor in the bay – if all had gone well with her. Then came thought of the fearful predicament that his nonappearance presented to Captain Davis. He had left him authority to assume command of the expedition, and at what date would Davis have to write off Mawson and his companions as lost and dead?

They had planned to have everyone away from Cape Denison by 15 January. Wild had food for only one year, and had to be relieved or he and his men were dead.

How long could Davis wait for Mawson to appear without risking the lives of Wild and his men? Most certainly the shrewd, cool-headed Davis would send out a search party to seek some sign… but in what direction would they come? And how far?

The answer rested on the same factor that ruled his own chances – the weather.

The contemplation left him more deeply restless to be moving; it renewed his intense preoccupation with the matter of weight on the sledge. He still had some miles to go before he could break free from the glacier. Then he had to scale the heights of the ice-dome to make the long crossing toward Aladdin's Cave. Every ounce on the sledge would tell against him in that fight for distance. The last few scraps of dog meat could be cooked so he would pull less weight in fuel; there were old, worn socks to be discarded with some underwear in his yellow ditty-bag – the Paquita bag, Ninnis had called it. Once clear of the crevasses he could throw away his rope ladder and the spiked crampons he'd worn over his fur boots; there should be steel-spiked crampons depoted at Aladdin's Cave – if he should get that far – which he would need to make the long climb down to the hut… and ever and anon, he noted in his diary, thoughts and facts came back to food. He would have to eat better, fortify himself against flagging on the trail with the 'perks', the raisins and almonds and the extra sticks of chocolate he'd saved for the toughest days. He would also dip into future rations. It was all pressure on his ailing body and his mind; and he woke at odd hours during the windy night hoping for signs of better conditions that would let him climb free of the glacier.

He roused from intermittent sleep at 6 am. His body felt bruised and strained. The accidents in the previous days, the crushing struggle from the crevasses, the toil of fashioning the rope ladder, the efforts and exertions that had followed were, he thought, the reasons for the distress, nausea, and faintness. He could not stand upright, and bent forward, canted to the left, to ease the pains in his right side. When he heated his can of dried pemmican powder that morning hairs from his bearded face fell into the mixture, along with fragments of scaling skin. He felt ill, and he ascribed this to weakness from starvation and toil, but there was at the back of his mind a condition fearful to polar travellers, a word that he would not then consider, even though some of the symptoms were already plain on his body. He would not allow the idea to have place in his mind and he would not even write the word – scurvy – into his diary... not yet!

He had a fearful solitary journey still to make across a vast plain of ice with violent winds, on rations barely enough to support life, and any added anxiety would affect his physical response. He had to keep hope alive, for that was to keep himself alive.

The weather on the lower plateau held against him. On this morning of 20 January, he sat in his tent idling the hours away while the torrent of air filled the bowl of space about him with huge flakes of madly whirling, veering snow, and in despair he went down on his knees and prayed for a break in the constant obliteration. He half-packed his sledge, assembled his tent gear for quick loading; and he waited, hour after hour. He reached agitated frustration: 'At 2 pm I started off in desperation without seeing anything of the land underfoot.' Boldness was rewarded; conditions lifted a little, the sledge ran well, and he made a further two and a half miles before the rising gusts blew so strongly in his face, coating his goggles with snow, that he was forced to halt and put up his tent.

This was the pattern of his subsequent days, a grinding fight uphill, the wind growing more boisterous with altitude, He was advancing westward, and so he was moving deeper into the true weather pattern of Adelie Land.

–

The days grew more harrowing, the winds stronger and the slopes steeper. Each night at the end of his daily attempts to make ground he wrote his diary entry: 'Jan. 21… I got started about 10 am, over a surface of deep snow and straight away had to tackle a steep slope – steeper than any slope from winter quarters to Aladdin's Cave. Then there were more slopes – and they continue ahead. *Very* heavy work! Later the snow got wet or I should have done more than the two and a half miles. I leave crevassed ice on either side. I threw away my crampons and the spare alpine rope. There seems no hope of a fair wind here – the winds all blow against me.' He was so concerned with weight he even ripped the cardboard covers from the diaries he carried and threw away the unused pages.

The next days were little different. On the evening of Wednesday 22 January, he was getting higher on the ice canopy and saw a fine view to the north-east of the eastern side of Commonwealth Bay: 'Had I known how far south it came I would have made in that direction long ago... I have felt very weak after the march today and I must increase my ration, for the snow is so soft and so deep, and so hard to walk in.' He appeased his hunger by raiding the bag of food he had allotted for the next five days. He ate a good pemmican hoosh and a whole biscuit.

–

As he gained higher levels of the plateau, the wind and blown snow further impeded his progress. Strong gusts capsized the sledge several times, and when he came to polished ice he constantly staggered and fell, jolting his aching frame. On 24 January he could look back down the ice dome and see how far he had come, and there he saw the Mertz Glacier snaking from the inland icefalls to where it projected its mighty tongue out to sea. He could see Aurora Peak in the north, and how far south he was of the outward course. Seeing the land like this he reasoned the wind fell from the desolation of the inland polar plateau and, rolling over the ice-brow, dropped down to sweep along the glacier bed. That day he met those winds head on. When he had covered a mile or two, wind gusts blew him from his feet, turned his sledge over and filled his goggles with snow. He was under enormous strain holding the sledge to its course and felt drained of energy. 'Not certain how my strength would go under those conditions, I fortified myself with a stick of chocolate and carried on until it stopped a bit and then hurriedly got my tent up.' It took him several hours. The effort was so wearing he broke routine and had hot cocoa before his meal. Then he wrote his diary: 'Reckon to have done three and a half miles today, but drift and the unreliable compass make it difficult to tell. I shall have to continue the improved feeding notwithstanding the poor outlook.' Then, on a separate line:

'Both my hands have now shed skin in large sheets. It is a great nuisance and hands very tender.'

He was near the brow of the ice in this camp, and during the night the full viciousness of the wind assailed his fragile shelter. He described it as 'a violent blizzard with winds above sixty-mile strength'. Hopes of moving on were stripped away in the howling blast; hard pellets of snow, flung from the bitter uplands, shrieked into the tent fabric like frozen bullets; the tent frame rocked – until the snowfall crept higher and higher up the cover, stilling its movement.

All through the day the gales blustered and piled the snow against him. Disconsolate, he complained: 'Cannot move. Could do nothing by myself in this weak state in this wind – it would blow all my possessions away. Besides, there would be no hope of striking camp again. The tent is banging; I keep trying to think of all manner of things to while away the time. In the end I always come back to food, how to save oil; and I am going to experiment and heat the pemmican, cocoa, and the dog jelly, all together. My feet freezing; I have to wear my waterproofs in the sleeping-bag.'

The snow was creeping higher and higher up the side of his tent. Soon it was more than half-way and he became apprehensive. If it reached his peak, four feet above the surface where he lay – would he suffocate? Surely, he would be able to dig himself free for air. The tent frame was pitched narrowly. He could see the weight of the heavy snow pressing the fabric in all round him, forming to the shape of his recumbent body, shaping down from the shoulders to his feet.

'The tent has taken the shape of a coffin – a coffin of blown snow,' he noted. To his mind, it was looming death. If the blizzard lasted very long he would rest here, never to walk again. 'It makes me shudder to think of that at the moment,' he scribbled. 'I will remain full of hope and reliance on the great Providence which has so far pulled me through.' If he was stronger – even if there were two of them, he argued – then this wind might be used to blow the sledge along. 'But for one, it is out of the question.' All day long the wind bludgeoned down from the uplands to roar through the glacial valley; he could do no more than rest in his bag while the casket of snow built round him.

At that same time – some fifty miles to his west – three men, hauling a sledge up a steep slope of ice, steel crampons on their leather boots, fought the wind to reach Aladdin's Cave, on the first stage on their search for Dr Douglas Mawson and his two companions.

–

Captain John King Davis held his crisis meeting in the hut at winter quarters late on the afternoon of 24 January.

Troubled, grim-faced and the skin taut round his eyes, Davis told the fearing men: 'I cannot hold the ship beyond January 30th. A party must go out to look for Dr Mawson, to search as far as possible and be back here by then. It can't be left any longer! Wild must be picked up at Gauss Berg – 1,400 miles from here – since he had only enough food for a year. There are ample supplies here for a small party. So, if the search brings nothing, if Dr Mawson doesn't return by next Thursday – January 30th – I shall leave a small party here at the hut, to be relieved next December, and I will sail to relieve Wild and his men. To delay beyond then will risk having the *Aurora* trapped in the pack-ice.'

The search party was the young Sydney doctor, Archie McLean (he had gone with Madigan on the coastal survey beyond the Mertz Glacier), Frank Hurley – fresh back from the gruelling 700-mile slog to find the South Magnetic Pole – and cartographer Arthur Hodgeman, who had travelled with Bickerton on the deep western survey.

On their first leg, on 25 January, they battled the wind to reach the ice-hole, a stiff upward climb of five and a half miles, which took them more than six hours. The blizzard held them in Aladdin's Cave for 24 hours. Then, after midday on 26 January, they pushed eastward on steep icy slopes for five miles and were forced to camp in a fifty-mile wind to spend a miserable thirty-six hours, their bags and camp saturated with wet snow, fretting at the dwindling time they had to complete their search.

They dug out their sledge on the morning of 28 January, and slogged through the hours of the day, covering sixteen miles before camping, in a heavy snow-fall, there to rest for the night hours before turning back for the two-day race to the ship.

The morning of 29 January came with a strong wind blowing thick drift horizontally across the bleak wastes. The three men were depressed at their fruitless search. Frank Hurley climbed a snowy mound and scanned the eastern horizon with binoculars, peering into the drifting, mournful white curtain that shut out the distant land-scape. Where would we go, if we went on from here? He questioned; in this vast arc of land which direction would we take? There could be no answer.

He turned to his companions: 'Lord knows where they are, whether they're alive or dead! We ought to leave a cairn, perhaps, just in case the Doc does come this way.'

They quickly built a good mound of hewn snow, and in the top they stowed a bag of food, wrapped in waterproof cloth, with a tin containing a note written by McLean. Over this they stacked further snowblocks wrapped in black bunting so that the tower should show up in the white waste. For the last time Hurley scanned the murky horizon to the east – again he found it lifeless. Dejected, they donned the sledge harness and turned back to Aladdin's Cave and the waiting ship. It was then just after nine in the morning.

–

On the morning of 26 January, Douglas Mawson's tent was two thirds under snow. He noted 'radial pellets of snow' that bombarded his camp like hail; he felt helpless, unable to move across this terrible landscape, and filled his hours by doctoring his raw, inflamed feet and listening to the tent banging in the sixty-five-mile wind.

When noon brought a lessening in the tempo, he decided in his restlessness to 'try another experiment'… to break camp in a wind near gale-force and to push through the thick curtain of falling snow using the sail to blow him along. He was going out to risk his strength if the gale was still blowing when he was finally forced to make camp. The wind gave him trouble when the weight of the blown sledge flung him to the surface, but he plunged on mile after mile. He built a shield in his mind against time, against pain and peril; he called up the poetry he'd memorised and the Psalms – 'I lift up mine eyes to the hills…' Hour after hour plodding through the snow, dredging out words to shut out fear, to keep back the spectre of starvation that walked with him.

There were fewer than three pounds of food – a tin of pemmican, and the last of the scorched dog meat. He had eaten deeply into his 'perks' to help him fight up the icy terraces – and now there was little left for him, and little left of him. Still, he covered seven miles that day.

His weak staggering brought him to a panting halt; by the time he cut the snowblocks, lashed the supports together and fought the wind to slip the cover over the frame, it was near midnight. Shivering, he heated his thin pemmican, and noted: 'I am in a terrible mess! Everything is saturated with the snow. In a few hours it will all freeze solid.'

The wind and the snow continued through the next morning, and on into the afternoon. He used the time again to doctor his body. He was shocked afresh at his condition; a poor, desquamated, desiccated skeleton of a frame, from which the muscle tissue had vanished, wide patches raw with the friction of clothing and harness, inflamed in the groin, his black nails coming loose, his teeth shaky in their sockets, his jaws aching. And his hair! His beard was harvested by starvation; the hair was off his head in handfuls. 'I have lost so much hair that I rival my reindeer sleeping-bag, which is now moulting heavily. It is a race between us – who shall be bald first!' He could not possibly move that day. He consoled himself with fixing an approximate position by dead reckoning: 'Aladdin's Cave should be on a course between forty-five and fifty degrees west of north – and at a distance of about forty-three miles.' The ice-hole he and Ninnis and Madigan had gouged into the brow of the plateau was now a scintillating paradise to him. The ship itself was a concentrate of heaven.

On the morning of 28 January, he crouched in his tent under a driving snowstorm, chafing at inaction. His spirit seemed at its lowest ebb. He had been in this cold wilderness for almost twelve weeks; for more than six weeks he had toiled in hunger and illness – in the three weeks since Xavier Mertz died, he had suffered crushing loneliness. And being alone made camping and breaking camp more burdensome than hauling the sledge across this country; even to crawl on hands and knees was less laborious.

He railed against the weather; against the wind, the drift, the overcast sky. 'Not being able to see the sun, day after day, is most dispiriting. All my gear is saturated or frozen stiff. No matter how far I travel, it makes me feel dreadful. If only there would be a break in this weather.' In his tent he fell to his knees and prayed for better conditions. It made him feel better at once. Nothing to gain in idleness, he argued. 'I will risk another tussle with the wind; I will have another bout with the blizzard.'

He calculated his food supply. It amounted to little more than two pounds weight – and mostly it was scorched pieces of dog meat, the last few segments of stringy muscle; hardly a day's ration for normal sledging. He might spend a week struggling to Aladdin's Cave – much longer if the weather did not improve – and his capability would slide downhill very rapidly from his present condition. As well, he had come that day on the first signs of last year's ice waves, the sastrugi corrugations that would slow him down. Never mind; the evening was fine and better weather was promised: he rolled into his balding skin-bag with the comforting feeling that he was getting close to the hut. Whatever the conditions he would make his attempt.

The next morning brought sharp disappointment. Across the bleak Adelie Land wilderness, a forty-five-mile wind lifted a stream of masking drift, a river of snow that shut his visibility down to a few yards. He braved the wind and snow to toil at breaking camp. Soon his head was swimming, his eyes burning, his body limp, and he rested a time in the flying flakes before facing to the west. At mid-morning the wind was catching his sail with enough force to drag the sledge to one side. He sat himself aboard and it moved over the surface like a sail-boat; he steered it in the general direction, as best he could. After midday his horizon lifted to a few hundred yards; he was still moving slightly downhill. He stopped to nibble a stick of chocolate for his lunch, then went on, but now the wind fell away and he hauled his load into the snowfall, head down and his body bent with pain and effort.

Fighting the illness, subduing his hunger, Douglas Mawson trudged slowly into the snowy afternoon oblivious he was on the verge of the turning point in his struggle for survival. It dawned slowly into his awareness just after two o'clock when, through the wavering patterns of falling snowdrift, his tired eyes noticed a blur of dark colour, a black shapeless smudge; a smudge of black in a landscape that had only shifting shadows of white, mauve and lavender? He knew of no rocky prominence breaking through this frozen canopy, and, as he plodded the 300 yards to his right to investigate, he wondered what it might be. So he came on his thrilling, golden find, the moment that he called his 'miraculous, marvellous good fortune.'

The cairn was obviously freshly built – only a little snow had coated the top of the black bunting. He tore the top snowblocks away with frantic haste – and there was the waterproof bag of food, and a tin with a note inside. His thoughts sang inside him: 'Providence has guided me here! I could have come in over a sweep of a hundred miles. Yet, I'm here!'

He opened the tin with his sore, trembling hands and read Dr McLean's message – left in case the Doc 'passed this way' Dr Mawson had his first news of the outside world for almost three months. The cairn, said the note, was twenty-one miles, sixty degrees east of south from Aladdin's Cave; the *Aurora* was at anchor in Commonwealth Bay. All other parties had returned safely and the ship was waiting.

There was more news! Amundsen had reached the South Pole first, and Scott's expedition was remaining another year on the Ross Shelf. And then, the stunning information that at once had his head back, his moist eyes scanning the horizon to the west – vainly. The cairn had been built that morning; they had left to return to the ship about 9 am... and it meant that his camp last night had been no more than five miles from theirs. Instantly he wanted to rush after them! They could not be much more than five miles ahead! At the same time, he knew he could not hope to overtake three fit men hauling a single sledge between them. And here was the waterproof packet of food! Inside this cover the cache of provisions was contained in a bright red Paquita bag! There was ample food for profligate feasting – tins of pemmican, sugar, butter, biscuits, cocoa, chocolate sticks... and at the bottom, three oranges! The temptation at once to eat to gluttony was almost overwhelming... but the three 'dear companions' were only five miles or so ahead, and even if he had no chance of overhauling them it was his nature to try, to make the attempt.

The wind was from behind, so he let it fill his sail and push the sledge along as he sat aboard gorging biscuits and chocolate… and warning himself against the sickness that it would bring to his shrunken stomach. But – the oranges were a temptation not to be resisted by a man believing inside himself that he was in the grip of scurvy… they were symbols of another world, they told of orchards green and fruitful; they were miniature golden suns in his stained, dirty gloves. And the colour of the red bag! It brought a turmoil of emotion – his beloved Paquita had sat and stitched these bags: 'To brighten up your lives down there!' In the vastness of the frozen wasteland, looking at the coloured bag he felt suddenly very close to her, almost as though he could reach out and touch her. He was near to Aladdin's Cave and within striking distance of civilisation. But, the cruel land had not yet done with him.

TWELVE

Aladdin's Beacon

The little sail filled with wind. Its power pushed the sledge over the half-frozen surface at ambling pace and, since he could walk no faster, he sat aboard, revelling in his new-found treasure, anxiously scanning the trail ahead for signs of the three men – they might have camped for some reason; they might pause and look back.

He nibbled at chocolate and biscuit, and sat with the Paquita bag in his hand. If only they had come two or three miles further; had the blizzard not held him back for so long; and had the morning been clear and sunny they surely would have seen him. But this awful weather intervened, and he felt the full weight of disappointment. Being so near to human aid and yet so far diluted the excitement of knowing they were searching for him, and in his illness the surge of emotion caused him confusion.

He thought at first that the direction given in the note at the cairn meant he should travel back along a course that came sixty degrees south of east from Aladdin's Cave, and he had gone half-a-mile when he had second thoughts. It was probable, he reasoned, McLean had meant the Cave was thirty degrees west of north and so he now travelled a mile and a half in that direction. Then, again he changed his mind and thought he should go more directly north. The wind was less helpful in this direction so he hauled the sledge, bending in the harness, stopping at intervals to peer into the white murkiness, his heart aching for human contact – the sound of voices.

The miles came slowly under his plodding tread. Soon, the sloping terrain turned to hard exposed ice; he was among sastrugi, and ridges and sharply-fretted edges cut into his feet. The wind rose during the evening and tugged at his sledge, making him stagger. Soon he was falling every few steps, cursing the slippery ice, regretting the desire to lighten his sledge while on the Mertz Glacier that had led him to throw away his crampons. 'I had reckoned on getting to Aladdin's Cave without needing them again, and there picking up our steel-spiked crampons for the steep down climb to the hut.' Now he was crashing into the iron-faced ice, jarring his aching frame, the fear of a broken limb coming with every shattering impact.

It was evening, yet still his tired eyes scanned the hazed horizon for signs of the three men. It was empty, lifeless. Then, for a brief moment, the evening gale lifted and he saw open water to his north – Commonwealth Bay was down there, he could see icebergs floating in the water. He had come dangerously far north of his true course, and was now on the steeply sloping glacial ice of the coastal cliffs. He saw his peril at once. Blown northward by the constant tug of the wind he was at risk from a strong sudden gust that, in the haze and drift snow, could have taken him and his sledge over the ice cliffs to certain death hundreds of feet below on the frozen rocky shore. Now he had to face a grinding climb back to the ridge of the ice dome and his body took a pounding more painful than he had suffered on the Mertz Glacier. More than ever he felt the need for crampons – he could stand the pain and the jarring no more. He fell to his hands and knees and crawled back up the ice, dragging the sledge behind him until he found a patch of half-frozen snow. He reckoned he had covered eight miles from the cairn, that he had travelled thirteen miles in all since resting. And still no sign of Aladdin's Cave. His pulse was pounding, his head was spinning with weakness; he had to camp, to fight the wind, to lash the ski shoes to the splinted theodolite legs, to chop out chunks of snow for the skirt, to spread the cover over the frame.

It was near midnight before the blue flame of the roaring primus lifted him from melancholy: thick, hot hoosh, butter, biscuit and cocoa with dried milk – food, wonderful food… and no more dog meat; never, never again! The feeling of being full for the first time since Ninnis died was his only comfort. He was puzzled, confused, still doubting as he wrote his notes that night:

> 'Did McLean give the correct bearing? It
> almost seems to me that I have come too
> far west. Aladdin's Cave must be near – but
> I cannot see. Am now on ice and half-ice
> and keep falling every few yards on account
> of wind. If a calm day tomorrow I can do it
> – otherwise not, as I have no crampons and
> keep falling. It is a great joy to have plenty

of food, but I must see I do not overload –
or disaster may result... What a pity that I
did not catch McLean's party at the cairn this
morning!'

He wore his windproofs in the balding reindeer sack,
to increase the warmth of the food. His thoughts were
busy on his problem of crossing the steep wind-polished
ice without injury from falling. Once more his fertile
mind found an improvisation. In the morning he would
fabricate his own crampons. Also, he would throw away
the skin casts of his separated soles, which had now
hardened at the edges and were abrading his flesh.

He attacked the mahogany case that held the precious
theodolite. He took it apart with his little Bonzer knife.
Outside the morning was full of howling wind and
screaming drift, too vicious for him to stand on the glassy
slope; he was happy to busy his hands with his new project.
He cut two wood platforms to fit his feet and prised and
levered nails and screws from the box to drive through
these wood sandals to project underneath. It was slow,
painful work with his skinned hands; it took him until
early afternoon. Then, when the wind lifted higher from
over the canopy, he braved the weather to break camp.

His heart missed a beat when he saw in the clearing
air how close he had been to disaster the night before;
the sea was nearby, the cliff very close. For the first time
since leaving the hut on 10 November he could see the
ice-caked Mackellar Islands out in the bay. There was no
sign of the ship.

He would not allow himself to dwell on that just now. He had to win his way to the ice-hole, to rest, and from there to make his way down the only highway they had found between the hut and the frozen canopy. Aladdin's Cave was his target. He sat on the edge of his loaded sledge and bound his coolie-type wooden sandals to his feet with lengths of lamp wick, and then put on the harness. He had miles to go, up the rising ice, with the wind abeam to where he thought the beacon marking the ice-hole should still stand. The make-do crampons did well for the first few miles, but the wayward wind yanked his load away from his course and added to the strain on his feet. He drew solace from the thought of food but recognised it had come late, that it was still too early for it to strengthen him. Soon his skinny body was suddenly stabbing with new pain from his inflamed soles; he stopped, and on inspection found the impacting plod of his feet on the steely ice surface had pushed the nails and the screws back through the wood – some of them were driven up through his finnesko into the raw flesh. He rested for a while, knocked the nails and screws back into place, and then wrapped sacking around his feet before going on. His troubles were just renewing. He came to a region of badly crevassed ice: some of the glacial cracks were open and easily seen; some were compacted with half-frozen snow – and once more he knew the awful breathlessness of the ground vanishing beneath his legs... It happened several times. Once, only the spar he had rigged for his makeshift sail held the sledge from plunging to destruction. Wedged several feet down in the mouth of the yawning abyss, it brought him a mounting despair. He had no strength left to pull it free; he sat in the snow, his wooden soles on his feet, his head down, and waited until he could face the toil of unpacking the gear from the wedged load. Packages that three months before would have been lifted with a powerful arm were now ton weights. As he toiled on the slippery edge of the crevasse, panting and giddy from effort, his feet suddenly slid under this body. His crampons were split, broken apart, and – even worse – he felt the pain of stretched ligaments in his right ankle, and the burning ache of a jarred right knee.

He was now near total exhaustion. His sledge repacked, he limped on up the ice-dome seeking only a small stretch of flat packed snow to camp. When he came to the spot, he halted before starting to erect his tent. Overhead, the wind had lifted into the sky. Up higher on the dome, about a mile distant, he thought he could make out a straight dark line rising upward through the blur of the blown drift. Oh God! Was that the beacon? Was that the mast they had erected when they cut the ice-hole? Was he really so close to his citadel? Even so, he could go no further. It was near midnight and he must rest. He made only a short entry into his diary: 'It could well be the beacon. It was about west from me, but it seemed hardly likely to be so far as placed by McLean's bearing at the cairn. But, distances here are so deceptive.'

The blizzard was at full cry in the morning. He could not move. His leg pained and he sat and listened to the gale booming about his camp and flinging shutes of snow from the uplands out to the sea. He spent all day under the windy bombardment and again busied his hands on fashioning a new pair of modified crampons. Two layers of wood for each foot, with little crossbars. He prised more nails and screws from his cooker box and the sledge meter attachment, and on each crampon drove these through the lower platform. He was pinned to his patch of snow all day and night, not knowing if he had actually seen the beacon marking Aladdin's Cave – or whether his pain-hazed brain and his longing to reach his objective had fastened an image on his mind.

The morning of Saturday 1 February brought no deliverance. The ice-dome was still a bowl of white, whirling chaos; the tent slapped and battered against the tent props, warning of the treacherous wind scouring the slippery ice. He worked with his hands through the morning, trying to improve and strengthen his pitiful crampons. He cooked some food; he rested his injured leg, and tried to wait as patiently as he could for the wind to abate over this terrible land.

It came in the early evening – a failing away of wind and a rising of the snow stream into the higher air. He clambered from the tent funnel and stood straining his eyes into the west. And there it was… Aladdin's Beacon, still upright in the wire guys, but the flag long ago ripped and shredded by the wind! He went through the ritual of collapsing the tent and packing the sledge. He bound his new crampons to his feet and limped upward, across the ice. There was some sastrugi in his path, and the sledge rocked on the uneven surface. The improvised crampons did not last long; soon the boards split again under his feet.

He went on, for two and a half miles – and then he stood at the vertical shaft that led down to the ice-hole Ninnis had named Aladdin's Cave.

He tethered his sledge with the battered, lashed spade; he climbed down the shaft and, kicking the snow away from the door, crawled inside. There was nobody there, but it was heaven. For more than eighty days he had slept under the flapping fabric of a tent cover – now here were solid walls and a roof that shut out the wind and the noise.

Uplifted, excited at finding the cave, he rummaged through the material left scattered about the ice-floor. All the parties had come back through here, including the support teams, and there was evidence of men being in a hurry to get down to the comfort of the hut after being weeks out on the plateau. There was an opened box of biscuits, half-eaten tins of pemmican, discarded finnesko and socks, milk powder and cocoa tins. On an ice shelf he found Xavier's dog-eared copy of Sherlock Holmes... and there was memory of their night sleeping in this ice-hole and his talk of omens. In the litter was an old benzine case, pieces of wood, old food sacks and, of all things, a pineapple! But – there was nowhere a sign of what he so keenly sought.

Feverishly now, frantically, he searched the cave again, seeking the steel-spiked crampons he had left on the shelf almost three months before. They were gone. Vital to his hopes of a quick meal and then a trudge down the ice-slope to the hut, they had been taken away – and he was now left with his wooden soles, already breaking up.

For one wild moment he was tempted to recklessness; to break out, defy the lashing wind at his back and go down the treacherous iceway. But, his arms were weary, his legs weak; the make-do crampons would be near to useless on the exposed slopes. He sat in the cave and ate chunks of pemmican with a spoon, not waiting to cook it, and then, with his sledge still unpacked, he went outside to start his downhill trek. He was not yet used to the quiet in the ice cave; outside, the gale was more blustery, gusting snow over the polar plateau. He saw danger in his weakness and went back into the shelter of the cave, feeling dreadful disappointment. He consoled himself by arguing that it was already past seven in the evening and he could just as well wait till morning, to rest for the exertion that would be needed. Anyway, to arrive back on a Sunday would be fitting. He could hold a brief service and note it was on a Sunday they had set out on this tragic journey – he, and Ninnis, and poor Xavier.

They were with him in the cave that night, almost substantial in the warm yellow glow of the lantern on the walls, in the blue light of the primus, in the glinting crystals of his breath falling on his stained, shabby clothing. And in the midst of this sorrow, and his dashed hopes at not getting down to winter quarters that night, he still felt utter gratitude for his deliverance:

> 'Found Aladdin's Cave; great joy and thanks-
> giving! Praise be to the great Providence who
> has brought me safely here. My new cram-
> pons want improving; one is quite useless and
> I have strained my leg...'

He stretched out his moulted reindeer bag on the pile of old food sacks and waited impatiently for the hours to pass. Every now and again he clambered out of the shaft to look southward, and always the blizzard wind carried veiling clouds of snow. He comforted himself with the thought that never again would he suffer the ordeal of pitching that make-do tent.

THIRTEEN

A Winter to Wait

The blizzard rampaged above the ice-hole all night long.
Through the night and morning Douglas Mawson made
frequent sorties, poking his way through the snow-wall
the drift built at the end of the entrance shaft, scanning
the outside scene of chaos for a sign of hope. At 11 am,
that Sunday morning he crawled, dejected, back to his
skin bag to write:

> 'Far from abating, the blizzard continues with
> great fury – so, I must camp in this comfort-
> able cave.'

Just the same, he wore his trail clothes and windproof
jacket and kept his precious possessions assembled ready
for a quick start when the awaited lull came. His sledge
stood tethered outside, but he needed new crampons. In
the afternoon the blizzard rose to reach hurricane force.
It seemed to unleash its worst assault that day to crush his
optimism, to keep him entrapped in the ice cave on the
brow of the plateau, to deter him from escaping down
to the coast. In order to repel the malaise that attacked

him in idleness he kept active in the sunken sanctuary; he cooked crushed biscuit in powdered milk but found it disappointing – to his muted senses the milk tasted stale; he started work on fashioning a new crampon for his damaged right foot. He did not think he had enough materials in the cave to cut out and build a pair. The new wood platform came from an old benzine case and he extracted more nails and screws from his cooker box and the theodolite case. He used a food bag to pad the sole, to ease the pressure on his raw foot and injured ankle. The work was painstaking, tedious, with his little hand-tool – it took until midnight to complete, but he was glad to keep his hands active.

He woke every few hours in the night, but always the world was awash with snow and filled with wind. He cheered himself by reflecting on how far he had come, what he had overcome. And without 'the marvellous luck of finding the lonely cairn in the white drifting murk' he might well have died in his tracks when the last handful of food and dog meat had gone. He came to the crampons again and again; it developed into his crampon crisis. Oh, why had he thrown those crampons away on the Mertz Glacier? With these makeshift adaptations, bristle though they might with nails and screws, he would not dare to go down the slopes when the wind blew with any strength above a breeze. As to food he was grateful, joyous to have enough to eat, but he criticised its freshness – the pemmican was eaten without cooking, straight from the tin with a spoon in large, cold mouthfuls – the milk and the biscuits were old and stale and he questioned whether the food was doing him any good. His feeling of illness was increasing, not receding. The sickness was like a tide washing over his mind and his iron discipline of the long march was waning. No longer could he force himself into movement, or keep his limbs in action; no longer did the jolt and jar of the trail dull the other pains and stabs in his dilapidated body. In the cave his deterioration grew plainer, every hour. Then, he forced himself to make a detailed inspection and to look carefully at the ruin that the awesome journey had inflicted on him.

'Oh! It has surpassed my utmost hope to have got this far but I am in a serious state.' His legs were badly swollen at the joints, puffed up with fluid; his battered feet were inflamed, covered with the most fragile new skin, which had broken and was oozing pus. All his nails, feet and hands were black, some loose, all cracked about the bloated tips and festering – he thought they looked like the hands of a decomposing corpse. The tissues of his face were deplorable; fissures lined his mouth and circled his nose and eyes, unhealing cracks that weeped. His muscles had wasted away, and in his stomach and abdomen he bore the constant pains that made him write: 'My internal arrangements have all gone awry.' At long last he came to the dreaded word thrust out of his mind in the fearful fight back from the calamity at Black Crevasse:

'It almost appears as if scurvy, or something
of the kind is upon me! My joints are very
sore and aching; blood keeps coming from
my right nostril in a watery composition and

from the outbursts on my fingers. I am quite
sure the food here is stale – I may be getting
harm from it.'

The depressive effect of excess vitamin A was to stay on
him through these bitter days of delay and frustration, and
it brought to mind what the dreaded scurvy had inflicted
on the powerful frame of Ernest Shackleton on his first
long haul toward the South Pole with Scott. It was a
thought to close a long, dismal, doleful day – but he
ate some pineapple, half-frozen as it was, before going to
sleep.

–

Next morning the blizzard still stormed over his perch
on the brow of the icy canopy. He saw it charging the
snowclouds down the slopes and remembered how, at this
time the previous year, they had struggled to complete
the building of the huts. From the bayside they often had
respite when these gales still lashed over their heads into
the outer region of the bay, continuous gales that raged at
the altitude of Aladdin's Cave. He was in the teeth of those
falling katabatic winds that the polar regions flung northward, and deep, deep inland on the remote and forlorn
plains of the so-called Pole of Inaccessibility – which stood
in rarefied air 12,000 feet above sea level, at the height of
Erebus itself – the cold was intense, inhumanly low, so
that the air was always falling, rushing downward toward
the sea, toward the north… how long would it last? This
infernal gale-force blast!

To his eye the weather already had the wicked loom of winter. Since midsummer, that day six weeks ago when he and Xavier had shared the supreme delight of biscuit and butter, the sun's position in the sky had been slipping northward. The angled planet was now outward bound from the sun, on that section of its elliptical orbit that would take the bottom of the world to its most distant, dark point from the solar rays, to when the cold of the inland polar plains would plunge to unbearable depths, to a deep-freeze not yet measured by man.

The ominous signs were already in the sky – dark bands of threat from the south that had marked winter months at the hut. The crampon issue burned in his thwarted desire to break out of the cave – the distance was a feverish, nagging point… only five miles or so at the end of this interminable crawl across the ice. He tried not to allow the frustration to disconcert him – but he lived with it every waking moment. The details of the descent he would have to traverse were no longer clear but were blurred by the months of marching more than 600 miles across some of the worst frozen terrain then known in the world. Yes, there were long, gale-swept slopes of glassy, polished ice; there were declines rippled with ice waves and frozen weirs, ice-falls in which a man could quickly snap a limb; there were places where the pitfalls lurked close either side of the track marked by the mile-posts; there were jumbled upheavals to be avoided and old crevasses now choked with blown snow, and new chasms, surely, opened by the movement of the shifting ice-cap during the summer months. There was danger enough, but – there was no other route, no safer way. And on that last stretch of exposed trail, from the three-mile marker, to be caught in wind without efficient crampons, without strength to hold the sledge, a man was quickly dead. Wind – and crampons – these were the issues! He built added strength into his platform soles and then fitted a wider cross-bar to his sledge, an anti-crevasse device that might hold his load above a broken snow-bridge, hopefully. Yet, still he was blizzard-bound. Merging, and losing distinction, the days and nights of this immobilised solitude seemed, in his affliction, to be a cloud of time, filled with increasing lassitude and desperation, moving him always toward his final anguish.

In mid-week he was able to scramble from the ice-hole when the wind fell to about forty-mile force; he found his sledge overturned, the packages and gear blown into the snow down the hill. He recovered Mertz's ski-shoes – though he had no need of them – and repacked the sledge, lashing the load firmly and pinning it close to the cave entrance. As he worked, he kicked a few wood boards under the snow cover and, retrieving them, found they had nails at each end. Immediately he went back to the cave to work on a new crampon for his left foot.

His illness increased. He had to fight for self-control, and the anxiety crept into his notes, now growing shorter as the nights passed. On his fourth night in the cave he wrote: 'Blowing as hard as ever. I wait in the bag all day and listen for a cessation of the wind. Thick drift blowing in the evening. Oh, for a clear spell! Will the ship wait?'

By Friday he was too dejected to write more than six words – the shortest entry he had made since leaving on the far eastern journey three months before: 'Wind continues; too strong for crampons.'

It was his seventh night in the ice-hole... he felt defeated, trapped, forbidden by the awesome power of the polar plateau winds to go down, to cross the last icy bridge on this long terrible journey, to reach winter quarters and human kind.

In the early hours of Saturday morning the night winds increased their violence, and the gushing of the windstream rose to a roar. He again fell asleep to the sound, as though he was aboard an express train hurtling through the night.

Then, at 5 am, he had his first hint that the interminable delay might be ending. The diminishing batter of the gale woke him from fitful slumber; he sat bolt upright in the dark, cold cave listening with eager ears. No longer able to sleep, he cooked his breakfast cocoa, and with trembling hands dressed in his outdoor clothing, wrapping up his feet, packing his diaries and map notes, stringing his gloves about his neck with lampwick. When he broke through the wall of snowdrift to the outside, the sun was peeping through a sky of high, racing cloud. The wind was down to about thirty-five-mile force and he was certain this was his day. Come what may, he would go down! If necessary, he would ride the sledge and slide and sail, taking the risks with his anti-crevasse cross-bar, and his ladder... and his faith in Providence. By midday his slow painful preparations were nearly done; the wind had lessened and the light was good.

Still he waited, to be sure. The late afternoon generally brought a falling off in the plateau wind. He had noticed that in his march back and he would be certain to need every advantage at the end of a few hours struggling on the gradients with the sledge... he stopped to wrap rope round and round the runners at this thought – the grip would be needed on the steep polished ice. And it was after 1 pm when he started out. His new crampons seemed strong and firm under his feet, but his knees wavered and he trod cautiously.

He towed his sledge through the snow cover, into the first mile. His body creaked from eight days free of the iron resolve to move, his joints ached for want of the continuous rigorous exercise, but not once did he question the need to haul his pitiful load down this last tortuous trail. It was part of him, his symbol of survival – without it he would have felt naked and vulnerable. His thoughts as he shambled the first thousand steps were a sea of changing emotions: hope, joy, suspicion and apprehension. His actions were imposed caution; shuffle the feet, shamble forward and avoid impacting the bristling nails and screws too hard into the ice... already he could feel the heads of his make-do spikes driving upward into his padded feet. But – the thrill of having come so far! Of being so near! Head down, bent into the harness, his eyes straining through the glass of the old trail-worn goggles, he followed a dangerously wavering course with a turmoil in his head: 'To be here, after a terrible journey when all seemed so utterly impossible – Oh! It is so much more than I could expect. Always I intended to push on, to go ahead until the last, to reach some point where my remains might be found by a relief expedition, and our story told… yet, always inside myself there was hope against hope, the deep urge to win more ground… and now I am near those utmost hopes.'

He was keyed-up, anticipating deliverance, but he still feared the lash of the cold tongue of the polar plateau and shambled his way down in apprehension. The first of the mile posts came like a blessing; he was a thousand feet lower now and the wind was lessening. He walked toward his second mile post with uplifted heart; the drift was blowing above him and he stumbled his way onto smooth ice in a shower of gentle flakes. The sledge now menaced his safety – sliding and slipping, it threw new strain on his crampons and he felt the points of nails bending. He halted for the first time to lash more rope round and round the runners, to stop the crabwise or sudden forward movement of his load.

He could not stay longer to rest. He passed the two-mile post in a cloud of expectation and conjecture; soon, he remembered, when he came to the great roll of ice near the three-mile post, he would see a sweep of the bay – if the air was clear. What will be my fate? 'Has the ship gone? If so – have they left a party at the hut? Have they abandoned me altogether?'

He was much weaker than he allowed himself to believe. In the next hour his progress was a faltering shuffle, each step shorter, his gait slowing. Thus he shambled to the hill of ice where the surface canted below him, and here for the first time he could see the anchorage. He was about a mile and a half from the hut now. He pushed back his goggles and his eager eyes scanned the waters. The ship was not in her usual anchorage. Instantly he had the consoling thought – perhaps Davis has taken her eastward along the coast to look for us. But, far out on the horizon, beyond the mouth of Commonwealth Bay, was a black speck – a plume of dark smoke. A ship – sailing westward. And there was only one ship to be expected in these waters. Was he now marooned alone in this awful land?

Somehow, in the great loneliness he had endured in his frightful predicament, sick and starved to weakness as he was, he found strength in philosophy. So – the ship had gone. Whether he was alone or not – he could survive at the hut. 'What does it matter?' he asked himself. 'This terrible chapter of my life is coming to a close.'

A half-mile further on he stopped in his tracks when, above a rising ridge of ice, the whole surroundings of the hut burst into his view. With the scene the last strength seemed to flow from him into the snow. Slowly he scanned the surrounding apron; there was no sign of anyone, no smoke from the stove pipe above the white-coated roof. Then, with a leaping pulse, he saw three figures to one side of the boat harbour, working together, bent over some object on the ground.

In an eternity of seconds, he stripped off one glove and waved it with a tired arm above his head. It went unseen; he tried to call out but his voice was a croaking rasp, muted by his dried membranes and the clutch of emotion at seeing human beings.

Again he waved his glove. The seconds passed – then, as in a dream, he saw one figure straighten up and look toward the rising ice-cap. Dear, dear fellows – they still watched anxiously for him. The figures suddenly were galvanised into a rush of action, and soon the excited voices rolled to his ears up the steep, sloping ice. There was no more to do, nothing now he need undertake. He had reached civilisation; he was filled with utter gratitude to Providence for such a deliverance. Now the outlook for him was so transformed, his thanksgiving brought a flood of warmth and the release dissolved his tension. For the first time since Ninnis died the onus of the long struggle and the privation and illness overtook him. He fell against the sledge.

It seemed an aeon of time before the head of the first man appeared over the brow of the ice ridge. It was a face at first indistinct, and shrouded in a woollen balaclava; but the man started running on seeing him by the sledge, and soon he could see it was Bickerton, Frank Bickerton – good old Bick!

He reached Mawson and bent over him, anguish and sympathy in his face when he saw the emaciated wretch in the ragged, stained clothing; he put his hands under the skinny armpits and easily lifted the near-skeleton – now down to little more than a hundred pounds – and propped him onto the sledge. He knocked away the ice formed around the opening of the waterproof helmet, peered into the sunken eyes, the fissured face, wrinkled and skinned like an old walnut, and was aghast.

'My God!' he burst out. 'Which one are you?'

His head spun, his thoughts jumbled with the sound of voices, five of them round him now. There was Madigan – big and brawny as he once was himself – Bob Bage, Hodgeman, Dr Archie McLean – but at first they did not understand it was agony for him to speak. They were anxious, pleading to know what had happened. Where were the others? He tried to clear his thoughts in this most wonderful moment, then – hesitant, croaking out his story – he told the awful tale of how Ninnis had fallen to his death, and of the tragic end for Mertz. They did not need to hear details of his own ordeal. The tears were free on their cheeks and his own eyes flooded with the telling.

The moment held a dream-like quality, which clouded his memory and dulled the sharpness of the tragedies. It seemed, suddenly, to have happened long ago, in a different world. These men about him were real, very real, and instantly his heart and mind warmed toward them with affection. Dear companions, they had volunteered to stay behind in case he needed their help. They had turned their backs on the chance to sail home on the ship... had he actually seen that speck of black, and the plume of smoke on the horizon?

'The ship? Where is the ship? Is she here or did I really see her out there?'

He saw his answer in their faces. Bob Bage put a kindly arm round his skinny shoulders, his eyes serious: 'She has left to pick up Wild, D.I.' Then softening the blow: 'Davis wanted to get away at the beginning of the month but was held back all week by a fiend of a blizzard; couldn't get the motor-boat in to take off the main party until this morning. She sailed about six hours ago.'

It did not then strike him hard. The crushing burdens of his 300-miles fight for survival had blunted his reaction. Supporting him, taking his sledge in tow, they helped him down the last testing ice slope and, half-awake, he was soon in the hut, laid out on his bunk, feeling the warmth of comradeship. Bickerton heated water to wash his dilapidated body and McLean pored over him with ointments and creams. He touched Mawson's sore places with tender fingers, peered into his eyes: 'You are going to take a long, long time to climb back. You are badly in need of nutrition. I was putting a shoulder of mutton into the oven when they called that you were coming. Would you like some for dinner – with boiled potatoes and peas?' He blanched at the thought of the fatty food; his organs, his stomach, could not face the enormity of a roast dinner. He felt utterly indifferent to all but the most simple, the most appealing food: 'Oh no, thank you. All I really want is bread, butter and jam – and hot, fresh tea.'

He was enfolded in blankets; the hut was warm, he was dreadfully tired and the world outside was quiet, the evening strangely calm. But, he could not yet rest. He clambered to his feet and walked to the doorway. The men had already started their preparations for another winter in this home, readying for the thunder and bombardment of the winter's gales. Canvas and sacking had been stretched over the wooden frame to shut out draughts, and there...

on the rising land, rose the tall mast of the guyed aerial. It stood 120 feet high! They told him Davis had ordered the ship's crew to work with the base party to erect the mast, with an umbrella aerial at the top.

Wondering, hopeful, he asked: 'Does it mean we can call them up?'

Yes, Hannan had a receiver aboard the *Aurora*. They had made arrangements for him to listen every hour from eight each evening until 1 am – in case there was any news. He could not wait. At his desk in his little room he wrote the message: 'Mawson returned. Ninnis and Mertz both dead. Return at once to pick up party. Mawson.'

There was a new wireless operator at the hut, Jeffryes, recruited by Davis in Australia to relieve the work of Hannan during the voyage home. He now bent over the Morse key, tapping out the message repeatedly. They all stood and watched as if expecting the miracle of a reply. There could be none. There was only a receiver aboard the *Aurora* and, though it seemed to go well, and there was every reason why the message should be pulsing out from the tall aerial mast with enough power to reach the westering vessel, they could not be certain the news was heard. They would have to wait until tomorrow. She could not steam back to the bay until mid-morning, even if the message was received and deciphered at once. They tried to make more certain, however, and Jeffryes was kept at the key until the closedown sign at 1 am. If the ship was ever to be recalled to pick them up, she would already be on the way back.

They had time to gather personal effects and pack essential items and to sleep a few hours before the time of embarkation. There was no luxury of sleep for Douglas Mawson. The feel of the bunk mattress was unreal under his body, the blankets were strange and disturbing; the excitement, gratitude at his escape, the thought of future living filled the hours. It was to be weeks before he could again sleep through a night – and months before normality returned to his internal processes.

–

The *Aurora* was a quiet, sad ship. In the Antarctic night she butted the floating pack-ice on her long westward journey. The men of the relieved main base party were in their bunks, most of them still recovering from the hard weeks of sledging – Webb and Hurley, especially, from their marathon 700-mile march to the South Magnetic Pole region, fighting the southerly gales at 7,000 feet on the inland plateau. And there was more than aching weariness on them – there was a deep-laid grief of loss. It had been with them for the last two weeks. 'A week overdue means they are in dire trouble,' Webb had said. 'To be three weeks late in that terrain is cause for the most serious doubts.' Gloomy Davis had waited as long as he could, and then came the returning search party. 'No sign anywhere!' The fear that the far-eastern party were dead was now open. The departure of the ship, to fight through the pack-ice and take off Wild and his men, could be delayed no longer. The awful, week-long blow then intervened; it raised great seas in the bay and they watched from the hut as Davis fought the gales to save the ship, losing five anchors in the struggle not to be smashed on the reefs – steaming up and down, day after day, consuming precious coal. And then, with the blast down to about forty-mile force that morning, they had tossed in the motor boat from the shore – a moving, painful farewell of the men left at the hut... a departure unlike anything they had dreamed of for this day.

The ship was now pushing west at six knots, fighting for time to get Wild off the Shackleton Ice Shelf – to get out of the dangerous pack-ice and back across the stormy Southern Ocean. They were all ready for sleep that night – until the disturbance. It began as a buzz – then crew men were coming and rousing them, calling them all to the wardroom. It was after 11 and Gloomy Davis stood with a slip of paper in his hand.

'Hannan has picked up a wireless message from winter quarters,' he said. 'I think you had better all hear.' He read the brief, solemnly, and added: 'I have already obeyed the leader's order. The ship has gone about and we are on our way back to Commonwealth Bay. I expect to get there by mid-morning. With any luck, we'll have them all off by early afternoon and will then resume this journey.'

Through the night hours hope was strong and warm in Douglas Mawson; but, when dawn came, hard snow blew down from the plateau and wind began to fill the silent world. He did not leave the hut, but he could see the sky was dark over the entrance to the bay. So began a long morning with all their dreams on tenterhooks, with anxious monitoring of the rise and fall of the mounting gale.

It was McLean who first saw the ship. He came rushing back from the rise they called Azimuth Point, white with blown snow, calling excitedly into the hut: 'I saw her! The ship has come back. She is pushing into the bay!'

The full blast of the gale came down to meet the *Aurora*. By midday it raged across the open water at a force above eighty miles an hour; it whipped the sea to angry waves, to spume and blown spindrift, then to seething turmoil. At times, the head-on buffeting defeated the thrust of the single, four-bladed propeller, and in consternation the men aboard saw her going backward, being blown into reverse, through her own wake. Ice built up in her rigging and snow shrouded her form; every hour was a gamble with destruction on the reefs and ice-coated islands. Up and down, they were seeking a lee, all day long, back and forth. Hour after hour the ship beat against the wind, some fifteen miles offshore, her furnaces gulping the precious coal and roaring at full blast to maintain the vital head of steam that shielded all their lives.

In such weather there was no hope of taking the party off from the hut; the motor-boat could only live for minutes. Thus, on ship and on shore, the two leaders faced the same tormenting predicament; how long would this new blizzard rage? How long could Captain Davis battle the wind and burn his fuel in Commonwealth Bay and still have the chance to take off Wild and his men from the Shackleton Ice Shelf – 1,400 miles to the west – and then break free from the pack-ice with enough coal to cross the ocean to Hobart?

In the blizzard-bound hut that Sunday evening, 9 February, Douglas Mawson concluded he was in no position to solve the dilemma. Discussing the predicament with his companions, he said: 'I don't like commanding him to remain. He is responsible for picking up Wild and his group from the west, and for the safety of the ship and crew. He has to judge when he must go and – having got in – when he can still get out and get the vessel back to Hobart. They have spent a very anxious and trying time hanging about here for weeks in bad weather – and I don't know what his coal position is like. I must leave it up to him.'

He wrote the message to Captain Davis and it was tapped out on the key several times before they went to their bunks: 'Am most anxious to get off. Hope you can wait a few days longer but cannot command you to do so and give you the option to decide whether or not to remain.'

–

The ship was no distance offshore, beating up and down in the gale. The signals were never heard. High above the frantic wind-tossed clouds of snow the Southern Lights gleamed into the night sky with magnetic forces that blotted out the pulses of Morse code. It made no difference. Captain Davis took his own decision – independently. Again, he summoned the men from the main base party to the ward-room where they stood with the decking heaving under their feet and saw the captain's bony face grim as he tersely pronounced:

> 'Time has run out. I am going west. There is only about enough coal to get the ship over to pick up Wild and then get out of the pack back to Hobart. I am doing this because I consider Wild to be in the greatest danger. I am responsible for him and his men and for the safety of this ship and its crew. The party at the hut have shelter and stores for the winter and there is a medical man on hand to tend to Doctor Mawson – whatever his needs might be. I will raise the additional funds and bring the ship south again next spring to relieve them. That is all I can do. There is no other choice.'

It was after midnight when the *Aurora* broke free from the stormy waters of Commonwealth Bay and pushed westward at her top speed of six knots.

–

On the morning of 10 February, the men at the hut woke to find the wild wind still flailing the face of Commonwealth Bay. As they watched the air cleared, and they could see the flying spume laying new frozen drapery over the myriad little islands. It was a scene of quick transformation; the gale lifted high above and the blown snow showers from the inland veiled the far western highland. Through the morning, as they watched, the gale fell to a stiff breeze, the flying cumulus raced to the east. By midday the overcast had gone, the wind fell away and a dead calm spread across the water, which rippled in the returning sun. There were deep, beautiful shadows of mauve and violet on the sedately dipping icebergs parading further out.

It made a majestic vista as the horizon sharpened in the clearing light, but it was without life. The ship had gone.

They came and told Mawson, disconsolate, and he tried to cheer them with the thought that Davis might yet find he had enough fuel to come back and take them off in March. They all knew it was a palliative, and at once set about the work of making the hut more secure and draught-proof and storing food in the snow-covered verandah against the long dark months under the eternal blizzards.

Mawson felt entirely drained. His legs were more swollen than when in Aladdin's Cave; his joints were more painful; his stomach was sore and tender, and he suffered sudden bowel disturbances. Still, he did not lie and rest. He followed the men in their work about the hut, not to talk with them, but needing to be near them, to feel the warmth of their companionship. He listened to them talking of the prospect of Gloomy Davis bringing the *Aurora* back in mid-March, and how they would then all bask in the Australian sun by the end of April. But he knew, they all knew, that so late in the season there was little chance that Davis would risk the ship being captured in the pack – that, before they could all go home, they had another winter to wait in this accursed land.

FOURTEEN

Sequel: The Contract Stands

Battling the gale, the *Aurora* cleared the stormy waters of
Commonwealth Bay, and by 2.30 am, on the morning
of Monday 10 February 1913, she was nudging through
pack-ice in the Dumont D'Urville Sea, on her long haul
of 1,400 sea miles to the Shackleton Ice Shelf to rescue
Frank Wild and his party. Precisely at the same time,
2,000 miles to the north-east, another polar vessel came
through the summer night like a ghost-ship to the little
New Zealand port of Oamaru, midway down the east
coast of South Island.

This vessel, a converted three-masted whaler, showed
no lights and flew no flag. The nightwatch at the port
entrance was puzzled. Instantly the signal flickered across
the calm dark water: 'What ship is that? Please identify.'

There was no reply; mystified he signalled again. The
darkened vessel dropped anchor in the roads and he saw
a dinghy put off. Two men rowed the boat ashore and
he met them at the landing jetty. After quiet explanation
the harbour-master was called from his bed and then he,
in turn, roused the telegraphist. It was 3 am, when, with
one of the two men from the ship dictating, the Morse

key in the telegraph office started chattering the message. In this way the three New Zealanders were the first men of the outside world to hear of the death of Captain Scott and his four companions in their struggle back from the South Pole.

The secret transmission concluded, the two men rowed back to the ship and the *Terra Nova* at once weighed anchor and dissolved into the soft southern night, as silent as when she arrived. The ship spent the next twenty-four hours sailing north to Lyttleton – from where Scott had sailed south late in 1910 – judiciously out of contact while their news crossed the world to the contributing newspapers to which Captain Scott had pledged first sighting of his despatches. In this way the sorrowing survivors kept faith for their leader, then eleven months dead on the desolate Ross Ice Shelf.

Editors in London and New York were gripped by the account of the finding of the frozen bodies, the diaries with all the chilling detail of the heroic march to the South Pole, the stunning shock of finding the fast-moving Amundsen team had beaten them by weeks, and the remorseless crushing of five lives, one by one, in the pitiless cold. None knew of the weakened skeleton of a man, in a snow-covered hut 2,000 miles south-west of Oamaru, uttering another tale of death and awesome struggle for survival to a wearied wireless operator.

Through the evening of 10 February into the small hours, the murmuring continued with the ceaseless dot-dash on the keyboard; the story to be heard either by the relay station on Macquarie Island or, direct, by the station on the Bluff near Hobart. Nothing got through.

The face of the sun had been active. Enormous eruptions, violent beyond imagination, had flung vast clouds from the solar surface into surrounding space, and energetic electrons and atomic particles, then not yet known to man, had fallen captive to the Earth's magnetic field. Highly charged solar fragments spiralled down to the southern axis of the planet, dispersing energy and setting the upper atmosphere aglow so that the Southern Lights – the *aurora australis* – shone with cold Antarctic beauty over those regions where recently Eric Webb, Bob Bage, Frank Hurley had come closer than any men before to the location of the South Magnetic Pole, and where Mawson and Mertz had faced the disaster of the loss of Ninnis in Black Crevasse.

Over all that wilderness the powerful electric disturbance flooded the sky and swallowed and obliterated the feeble pulses of Morse streaming out from the mast near the hut by Commonwealth Bay. Dismayed, they waited through the night for some signal to come back. The first long epic to be transmitted from the Antarctic was lost in the brilliant sky-fire.

Mawson wrote his story next day and Jeffryes tried again that night – and the next night, and the next. And always their only response was the crackle of the magnetic storm above their heads. In his illness, the nagging hours of tapping out the tale, his inability to send out messages he felt ought to go to the monarch saying how new lands had been named King George V Land, to sponsors and press links, brought nothing but further dejection. He felt they were talking into oblivion, that the hut was totally isolated from the world; he suspected the malevolence of the shining aurora overhead – but, as ever, he persisted. He ached to get word out to Paquita Delprat saying he was safe and would be coming home, but he could not do that; official business had to take precedence, and if he allowed himself the indulgence of a personal message there were six other men in the hut who could claim the same right... and obligation would not be met.

Dr McLean did his best to nurse him back to some strength and stability, even to devising an electrical scalp massager to try to stimulate growth to his hair. But, McLean was totally in the dark on the nature of his affliction. They discussed the peculiar facets of Mertz's death, his madness and his convulsions. They traced the deterioration as faithfully as they could, and McLean formed the opinion that Mertz had died of colitis. (Late in life, in letters to connections in Switzerland, Mawson showed he changed from this opinion. He said on medical advice he had come to the view that Xavier Mertz had succumbed to peritonitis following appendicitis. But he then gave no explanation to his own internal pain felt at the same time, as he never knew of the vitamin toxicity in the dog livers that they had both eaten.)

As the long, galling nights went by, with frustrating silence meeting all the transmissions, Mawson sank into a cloud of illness, shot through with pain alien to the effects of hunger and malnutrition. He told nobody. He wrote it into his notebook:

> 'My nerves! I find my nerves in a serious
> state. From the feeling I have in the base of
> my skull I am suspicious I may go off my
> rocker soon!'

He noted sharp pain in the side of his head, too, but he did
not link this with the suffering of Mertz in his last hours
when he held his ears, crying with pain. Mawson wrote:
'My nerves have evidently had a great shock – maybe too
much writing has brought this on. I must rest more.' He
turned again to his doctrine that busy hands heal sick
minds. He made a wood frame for his photograph of
Paquita Delprat and stood it by the side of his bunk; yet
his illness stayed with him, always thrusting itself into his
notes:

> 'I am an invalid. My legs have swollen very
> much. I am very much shaken to pieces... it
> will take some time to pull me up to anything
> like I was physically before that awful journey
> home.'

The days of dark isolation continued. The gales took a
hand; an aerial insulator broke, then one of the connecting
wires was blown down, and Bickerton scaled the mast in a
forty-mile bluster to repair the connection. Fourteen days
after his return to the hut they got through to Macquarie
Island, a historic broadcast of its kind across the Southern
Ocean. The written account of the journey was once
more tapped out – and Macquarie Island replied – but said
that only about one half of the message was intelligible.

Then came a piece of news that dropped like a bombshell into Mawson's convalescence.

'We have heard from the mainland station,' Macquarie Island reported, 'that Captain Scott reached the South Pole about a month after Amundsen but that he and his four comrades all died on the way back. The bodies of Scott and two companions, Wilson and Bowers, were found by search parties last November but the ship recently reached New Zealand with the details. They ran out of food and fuel.'

That fine man! He remembered Scott urging him to join him in the march to the Pole… 'We should be there together to raise the Union Jack – for the sake of the Empire, for all that we have done up to now. Come with me! I want you to share that moment…' Now he was gone, a corpse caught in the cold continent. The men around him in the hut saw the pain in his face, that his eyes were full with memories.

He told them: 'I am so sorry. I know what they must have suffered – I know what it means. I have been so near to it myself, recently.' His own suffering and loss rose again, vivid, overpowering, and he turned away from the shield of their companionship and concern. He went into his compartment to brood and reflect… like Scott, he had led fine men to their death.

He emerged into the general area of the hut about an hour later, his face thin, lined, ravaged still, but his eyes were smiling. Anxious, and mystified, his men watched him. He went to the store and collected a dozen penguin eggs. He cracked them into a big bowl, set three pans on the stove, melted butter into them. He said, not a word and the men respected his silence as being of special significance. Slowly he beat the eggs into foam and then, as they all watched, wondering, he poured a quarter of a bottle of whisky into the bowl and then cooked the mixture in the three steaming pans. Mawson served a portion for every man and then joined them at the mess table. His eyes swept their faces one by one.

'Gentlemen,' he announced, 'we are all about to eat for the first time a serving of Omelette Mawson – an original dish invented, and named, while in the field in the Antarctic, by our late and honoured companion from Switzerland, Dr Xavier Mertz.'

–

In the wireless transmission schedule next evening he gave priority to messages of condolence to the families of Belgrave Ninnis and Dr Mertz, and then sent his sympathy to the widow of Captain Scott. Still, though he ached for word from her, he denied himself the joy of sending news to Paquita Delprat, but he found calm and some ease thinking about his future life with her. He decided to design a house that they would build as their first home, somewhere on the outskirts of Adelaide.

He had another project. For this he enlisted the skills of Arthur Hodgeman. They chose a baulk of timber from the store and Hodgeman cut a cross, the arms and tip bound with metal bands. A plaque of wood was then cut from the side of Mertz's bunk, which was polished and oiled for bolting to the foot of the commemorative cross.

Mawson composed the inscription that Hodgeman cut meticulously into the wood with a hand-held chisel. It read:

Erected to Commemorate the Supreme
Sacrifice Made By:

Lieut. B. E. S. Ninnis and Dr X. Mertz – In
The Cause of Science. A.A.E. 1913.

When the cross was ready it was erected, with simple cere-mony, on the ice-bound rocks and against the backdrop of the great frozen plateau.

April came in with mad gales roaring and booming over the hut. They brought the first deep snows of the coming winter. Each day confined to the hut sharpened attention on the nightly wireless transmission; each man reached out with yearning to the outside, far-off world. On the night of 4 April, Mawson opened the schedule for personal messages – he included his own brief signal to his beloved:

'Deeply regret delay. Only just managed to
reach hut. Effects now gone but lost most
of my hair. You are free to consider your
contract but trust you will not abandon your
second-hand Douglas.'

He had her reply, written on the scratch pad by Jeffryes on the night of 6 April – 'the dearest draft of all that came in…' It read:

'Deeply thankful you are safe. Warmest welcome awaits your hairless return. Contract same as ever only more so.'

In his delight it did not dawn on him then that he had received the first marriage pledge ever transmitted by wireless to the earth's sixth continent.

Another message came through later in April that destroyed their hope that the ship might return to take them off before winter sealed the sea with ice. The message came from Hobart, from Captain Davis. The *Aurora* had reached Wild and his group, found them all well, and had taken them off on 23 February. Davis would try to raise more money for the voyage south to bring them home next spring.

–

Slowly, the Antarctic winter closed around the seven men; each day grew shorter as the gales rose in strength. Mawson occupied his mind with plans for a new scientific programme for this enforced stay and to a careful rebuilding of his health and strength. It was to be a longer process than he knew. The march back from Black Crevasse had come near to killing him. He realised this as the weeks of restless sleep went on and the pains continued to rack his body. So much so that, when talking to newsmen the following year, he ventured a rare opinion about his condition, and said: 'I know now the enforced isolation in the hut for a second winter in that land was a blessing in disguise. My state was so poor, I was so near death's door, I now believe that had I been in time to board the *Aurora* and sailed immediately I could not have survived the long, rough sea voyage home.'

However, the months in the hut brought new torments and troubles. The new wireless operator, Jeffryes, broke down under the stress of living in isolation in the long Antarctic winter. They were all appalled one night when Bickerton, who was able to read Morse, heard Jeffryes sending a signal telling Macquarie Island he was in fear for his life and was surrounded by six lunatics. When charged with this, he fronted Mawson and accused him of casting a spell over his mind.

Life dissolved into a nightmare for a time, keeping Jeffryes under observation and away from the keyboard, quietening his ravings, and trying to maintain wireless contact without causing undue worry at home. Mawson was moved to deep sympathy for a man not suited to the challenging isolation of the Antarctic, and Jeffryes was never to fully recover, even though he was later admitted to an institution.

In all these tribulations and through the endless gales and the dark days, Mawson longed to hear the cry he knew would go up when, in spring, the ship hove into sight and he would see again the long, thin finger of smoke on the northern horizon... 'How will we ever be able to express such a feeling?'

–

When that great event came, in mid-December 1913, he was still shaky, wan, very thin and hairless. On a day of pale sun, the *Aurora* left Commonwealth Bay for the last time and Mawson sat by the after-rail watching the receding plateau of ice. Moved by this moment of home-going, his head full of memories, he recalled lines from his favourite poet – Rudyard Kipling – and copied them into his notebook:

We bring no store of ingots
Of spice or precious stones
But what we have gathered
With sweat and aching bones.

It was the final entry in the diary of his Antarctic journey.

An Appreciation

Eric Norman Webb, D.S.O., M.C., C.Eng., F.I.C.E., M.E.I.C.

Last surviving member of Douglas Mawson's main base party.

This tribute to the late Sir Douglas Mawson was made by Mr Webb at his home in North Wales at the age of eighty-seven, some sixty-three years after his journey with two companions to the South Magnetic Pole.

Douglas Mawson is a giant figure in the history of the exploration of the southern continent. He led one of the greatest exploratory expeditions of all time in terms of territory covered on foot – a coastline of 2,000 miles, penetrated to a depth of nearly 400 miles by a handful of men – and made massive contributions to knowledge of our planet with resources that must be considered niggardly by modern standards.

However, it is for other reasons that I am happy to contribute this appreciation of the man who enriched my life, perhaps more than any other, by his personality, by his example and by his admirable leadership, and whose life-long friendship I prized above all others. I never had a stauncher friend, and, throughout his long life, he could stand and be compared with any explorer in history for courage, fortitude, endurance, resolve and loyalty to his fellows.

For these reasons, Douglas Mawson was always held in the highest regard among polar travellers, both in the age of heroics in Antarctica during the early decades of this century, and in more modern times. Yet, I fear his greatness as a man and as an explorer is too little known publicly, especially in Britain, merely because his feats were overshadowed by the Scott tragedy in 1912 and the Shackleton drama of 1914. There is today a strong need for a resurgence of interest, especially among young people, in his feats and fortitude and in the example of bravery that this book presents.

To me, when I was a young man in my early twenties, Mawson was already a hero. As a boy I had watched the Scott expeditions go but from my home town of Lyttleton, in New Zealand, in 1902 and in 1910. So, I was thrilled in July 1911, when I was invited to join Mawson's expedition as a magnetician. In preparation for this role I worked on the magnetic survey of Australia by the Carnegie Institution of Washington, D.C., USA, which took me to the outback late in August where, at the tiny railway halt of Farina in the north of South Australia, I first saw Douglas Mawson – quite by chance.

Mawson arrived at that station to join the train in which I was travelling. He came in a Model T Ford – widely known as a 'Tin Lizzy' – which was loaded with boxes of uranium samples (pitchblende) collected from the first deposits (which he had discovered and identified) in Australia. I knew he was in the district and at once recognised this tall, tough-looking man in bush clothes. He was rangy and very strong. I saw him bend down and lift those boxes, which were heavier than lead, and heave them with his long arms into the rail wagon. I at once introduced myself and, from then on until the end of his life, I knew him as a man always straightforward who would not put up with nonsense, and who at once attracted respect.

My admiration for him was confirmed when I joined the expedition and we went south to Commonwealth Bay. His leadership was unobtrusive, capable, and highly appropriate to the personnel from 'down under' and for such a scientific venture.

He was not a Shackleton, nor a Scott – but he was no ordinary leader. Shackleton had a magnetic personality of the kind that is physical rather than intellectual, while Scott was, in the main, a naval martinet with scientific leanings. Mawson was, above everything, an intellectual leader with utter motivation and selfless dedication to his objective, which he handed out to all of us in his party so that, by common consent, it became accepted and promoted as the policy of the expedition. Thus, when we saw how he was completely committed, so each one of us became committed to his own particular discipline. Mawson's dedication to scientific objectives infused a like spirit into us with the determination to emulate and excel the results of our peers. Other factors of great importance in this regard were his thorough and effective planning, his organisation, and his contagious enthusiasm.

I am confident that his motivation and dedication to objective was a prime element in his ability to endure the terrible journey back to winter quarters following the loss of Ninnis down the crevasse. This was a very great and awful shock to his mental equilibrium, but I believe that his objective and resolve were then more firmly fixed in his mind than ever. Mawson normally weighed fifteen stone (210 lbs) and this was all muscle and bone – no fat. When he got back to the hut, from what I heard he was skin and bone and weighed something less than eight stone.

To me, and I have had some experience of what he suffered in my own journey to the South Magnetic Pole, his survival was the outstanding feat of any lone traveller in polar history. I have no hesitation in saying that. It showed him to be a great man in many senses, but mostly in that he could face the ultimate challenge and not flinch. We have to remember that he had already spent a whole winter in the base hut under the most blistering gales ever experienced by men. In the month of May alone, the average wind velocity (the mean strength of the wind for the whole of the month) was more than sixty mph, which is something like six times the average for Europe. On many occasions, for several days together, the *average* wind velocity exceeded eighty mph, with gusts peaking 200 mph, and more. He was only too correct in saying that we had found 'the windiest place on the face of the Earth.' Moreover, he had marched with Ninnis and Mertz more than 300 miles over territory of the most unimaginable difficulty in uncertain weather and poor light. They met falls of snow that covered and masked the areas of ice ridged beyond belief by sastrugi. These are waves and ridges of compacted snow produced by strong wind laden with drift snow and are shaped like an upturned boat but upwind tailing away to nothing. They may be only a few inches high, or go on right up to six feet, sharp-ridged at the top running out to merge into the general plateau surface, hard and smooth as polished ebony – nearly solid ice.

The cracks or crevasses underfoot largely ran across their path, both going out and coming back. Mawson's sufferings from starvation must have been horrendous. On my own journey coming back from the South Magnetic Pole, I had a glimpse of what he endured when, already debilitated by inadequate rations and extremely arduous hauling, my party hauled the last seventy miles in three days on one fifth of the standard ration.

I was aboard the *Aurora* when he returned to the hut and thus did not see him again until late in 1914 when he was already married to Paquita and had been knighted by King George V. He came to talk with me on the scientific accomplishments of the expedition, the compilation of data, and analysis. The expedition's researches covered geo-magnetism, oceanography, meteorology, geology, biology, while the finding of traces of coal, copper, and valuable minerals had evoked considerable interest. It was a mountain of work calling for years of effort. Moreover, he, himself, had to undertake public lectures, travel, and the writing of his account of the whole expedition in his book *The Home of the Blizzard* in order to pay off the debts that had accumulated steeply because of the second winter he was compelled to spend in the Antarctic.

When we met in New Zealand, he was the same tough, self-reliant character I had first seen loading the boxes of uranium samples at Farina. He was still purposeful, but he was a noticeably chastened man – quieter, humble, and I think very much closer to his God. To me, he was a superman – the marvel that had survived – and he took it as a matter of course, but that was his modesty, underlaid by his faith. I asked him about this and he told me that there was some other power he had borrowed on that journey that was superior to the willpower that pulled him through. His faith steeled him; he drew his personal strength from this faith. Yet, I saw he had aged, was worn, had lost much of his hair, and I fear he was never again the same iron man who started on that fateful journey. I am now convinced his terrible sufferings left scars on his physique and his constitution, and that he would have lived a lot longer than he did but for his awesome ordeal.

One other factor that must be mentioned in relation to his escape from death on the ice was character. He had always led a frugal life and had learned in his early years in Australia to meet difficulties and rugged conditions and to put up with limitations. He was a marvellous innovator, and improviser, and always extremely resourceful in the most practical way. Perhaps these qualities led to the saving of his life; his innovation prompted him to choose alcohol as the priming spirit for our primus stoves instead of methylated spirits, and I know this was of much help to my party. He was, as I have said, a great man in many senses. Yet, the timing of the release of his story so close to the Scott tragedy meant it was almost completely eclipsed. This seems a wholly unfair quirk of fate, which put his great performance in the shadows. World reaction to the Scott epic obliterated public interest in Mawson and left him with a daunting struggle, when he did get back, to win the interest and support he needed to pay off the debts caused by the disastrous delay. I am afraid that this disinterest and ignorance of this great man still persists in many places, even in his homeland of Australia. I deplore this utterly, because his epic belongs not to any single country but to all men, as a triumph of the human spirit over the most formidable adversity.

Strangely, the Antarctic maintained its hold on his mind for the rest of his life, though he never again made a long sledging journey. Like me, he joined the Services during the First World War, and then took up again his most distinguished academic career in geology. By the mid-1920s the south was again calling him, and he sought to extend the complex operation of charting the coast of Antarctica westward from the Shackleton ice shelf, where Wild had landed in 1912, along the quadrant below the Indian Ocean into the area south of South Africa. He was unable to obtain the kind of modern ship he wanted and his choice fell on the old, but worthy, *Discovery*, the same ship in which Scott had sailed south in 1901. He asked me to join him on the voyages he made along this vast strip of territory in 1929-31, but I was unable to do so since I was already committed to civil engineering and involved in hydro-electric schemes in India and elsewhere.

His work, by these expeditions, established Australian interest in the Antarctic over a vast sweep of the Earth's surface. But, his contribution included an awareness of the resources and the wildlife in that region, which is only today bearing some fruit.

He has done much to promote a world-wide awareness of the plight of the greatest mammal on Earth, the whale. It is a world contribution arising from his anxiety over the plunder of the whale populations as early as 1911. He had great concern for all wildlife but, at that time, he became acutely conscious of the dangerous position in which man had placed these wonderful creatures. In later life, he never ceased his efforts to bring the plight of the whale to the attention of all nations and all peoples. Also, he protested ceaselessly against the senseless slaughter of penguins for their oil, which gave so little commercial return for the carnage that was inflicted.

For young generations today, the feat by this tremendous man, of surviving his epic, lone journey might have less appeal than the movement that he worked to set in motion, which was preservation of the unique wildlife in our oceans and our skies, so essential to our very existence. I have said he enriched my life; I hope that his story has like effect on many other people.

ERIC N. WEBB
Llandudno, N. Wales
1976

288

A Clinical Background

Modern medical textbooks and papers in learned journals have recognised 'Hypervitaminosis A' as a clinical entity. Put plainly, the condition is toxic reaction to an overdose of Vitamin A, the severity of which may be heightened by the absence of related Vitamins C and E.

Symptoms of Hypervitaminosis A are listed as follows:

Headaches, vomiting, and diarrhoea, all of which appear fairly promptly; scaling and stripping of skin – within a week – loss of hair, splitting of the skin round the mouth, nose, eyes. Then comes irritability, loss of appetite, drowsiness, vertigo, dizziness, skeletal pain, loss of weight, and internal disruption from swelling of liver and spleen, including violent dysentery. In severe cases this leads to convulsions and delirium and possible death from intercranial haemorrhage.

It has long been known that illness can follow the eating of the livers of polar bears, and clinical studies have shown that this danger extends to the bearded seal and, more recently, to the Greenland husky dog. Work in Adelaide reveals that a mere four ounces of husky dog liver is a toxic dose of Vitamin A for an adult man.

Greenland huskies and their viscera were eaten by Mertz and Mawson, but not for more than half a century after the journey described in this book was the link between their sufferings and the ingestion of the husky livers suggested. In fact, Sir Douglas Mawson died in 1958 unaware of this factor in his 'terrible journey', and, after brooding on the madness and death of Mertz, had been convinced it was due to peritonitis arising from a burst appendix. Most recently, vitamin studies have shown that scurvy effects from the lack of Vitamin C might be accentuated by an overloading of fat-soluble Vitamin A.

Author's Note

The explorations by the Australasian Antarctic Expedition of 1911-13 spanned some 2,000 miles of the unknown coastline of the world's last unmapped continent. As well, one party penetrated the inland to get closer to the South Magnetic Pole than anyone had been before. Details and results of the separate journeys were written by the group leaders and were published in 1915 by Douglas Mawson in the book, *The Home of the Blizzard*, which he wrote with the co-operation of Dr Archie MacLean. Readers interested in the whole work of the venture should consult Mawson's book since this work is mainly concerned with the tragic journey to the far east of the region, which became a trial of a human spirit.

Mawson's account of that journey has been the only writing on the epic during the last sixty years, and typically – and perhaps necessarily – modest, it was a tightly controlled narrative that precluded heroics. The new evidence, however, throws fresh light on Mawson's ordeal on the ice – facts that have only emerged since Mawson died (in 1958) make his survival even more monumental and remarkable.

In 1968, the late Sir John Cleland and his colleague, R. V. Southcott, first suggested that the death of Mertz and the illness of Mawson were due to Hypervitaminosis A. However, that the vector was the Greenland husky was not then suspected. In 1971, further work in Adelaide discovered high loadings of Vitamin A in the dog livers. There is normally enough in a single husky liver to provide ten toxic doses for an adult man.

The interest was whether the suffering and affliction of the Mawson journey could be matched against the known symptoms of Hypervitaminosis A, the clinical condition that results from excess of this powerful substance. In pursuit of this I found much in Mawson's sledging diary, which had never emerged into print through his own writing. It followed there could be more evidence in the unpublished diaries of the two men who had died on that journey – one of which, Mertz's, had been written in German. Mawson had not kept these diaries, and so I was launched on a search that took me across the world, and which proved as rewarding as it was remarkable. The original diaries were not located but valid copies were unearthed with the result that – along with a careful sifting of Mawson's own diaries – the bare bones of the harrowing journey could be found. All conversations in the book are based on entries in these diaries and recollections of close associates and relatives.

The last surviving member of Mawson's main base party, Mr Eric Webb, tells of Mawson in 1914 being a 'much chastened man'. He says he talked of a 'presence' that he felt guided him through suffering to survival and so became a man with strengthened faith in his God.

For many minds this could be reason enough for his noted lifelong reticence to talk of Mertz's death and his own subsequent ordeal, a reluctance that was breached only rarely and with close academic associates. There is, however, strong indication in Mawson's own correspondence with the Ninnis and Mertz families to suggest that he did not disclose in print the grimmer details of Mertz's ending because he wanted to protect the memory of a brave comrade. He did not need to soften the death. With the new evidence, Mertz's dementia claims only our warmest sympathy, and Mawson's own performance becomes even more impressive – and noble.

No other writing I have attempted has been so humbling as this incredible narrative, and if the work has any value then much must be due to those who gave so freely of their help. I must express my gratitude to Mawson's surviving daughters, to Mrs Jessica McEwin and her husband, Peter, and to Mrs Patricia Thomas. I owe a great debt to the knowledge and support of Dr Fred Jacka, Director of the Mawson Institute for Antarctic Research, in Adelaide, and to his secretary, Edna Sawyer for her patient sifting. At the famed Scott Polar Research Institute, in Cambridge, England, Dr Gordon de Q. Robin, the Director, and the Head Librarian, Mr Harry King, made their annals available; I also received help from the Australian Antarctic Division and from many people who knew Mawson later in life. I am also grateful to Mr Eric Webb and to Sir Edmund Hillary.

I have to acknowledge guidance on the medical aspects of the tale from Dr S. Nobile of the Roche group, Dr John

Duggan of the John Curtin School and Dr A. C. Kail on the effects of Hypervitaminosis A.

As well, access to documents and correspondence files in the Mitchell Library, Sydney, and the National Library, Canberra, have added much value to the narrative. Help with translation came from Mrs K. F. Jansen and with the preparation of the manuscript from Mrs Olga Moore – but the heaviest debt is to my wife: her help and immense forbearance made completion of the work possible.

LENNARD BICKEL

Cronulla, Australia

1976